Forecasting the
Environmental Fate
and Effects of Chemicals

Ecological and Environmental Toxicology Series

Series Editors

Jason M. Weeks
National Centre for
Environmental Toxicology,
WRC-NSF, UK

Sheila O'Hare
Scientific Editor
Hertfordshire, UK

Barnett Rattner
Patuxent Wildlife
Research Center,
Laurel, MD, USA

The fields of environmental toxicology, ecological toxicology and ecotoxicology are rapidly expanding areas of research within the international scientific community. This explosion of interest demands comprehensive and up-to-date information that is easily accessible both to professionals and to an increasing number of students with an interest in these subject areas.

Books in the series will cover a diverse range of relevant topics ranging from taxonomically based handbooks of ecotoxicology to current aspects of international regulatory affairs. Publications will serve the needs of undergraduate and postgraduate students, academics and professionals with an interest in these developing subject areas.

The Series Editors will be pleased to consider suggestions and proposals from prospective authors or editors in respect of books for future inclusion in the series.

Published titles in the series

Environmental Risk Harmonization:
Federal and State Approaches to Environmental Hazards in the USA
Edited by Michael A. Kamrin (ISBN 0 471 972657)

Handbook of Soil Invertebrate Toxicity Tests
Edited by Hans Løkke and Cornelius A. M. van Gestel (ISBN 0 471 971030)

Pollution Risk Assessment and Management:
A Structured Approach
Edited by Peter E. T. Douben (ISBN 0 471 972975)

Statistics in Ecotoxicology
Edited by Tim Sparks (ISBN 0 471 96851 X) CL
                      (ISBN 0 471 972 991) PR

Demography in Ecotoxicology
Edited by Jan E. Kammenga and Ryszard Laskowski (ISBN 0 471 490024)

Ecotoxicology of Wild Mammals
Edited by Richard Shore and Barnett Rattner (ISBN 0 471 974293)

Forthcoming titles in the series

Bioremediation and Land Reclamation:
Tools to Measure Success or Failure
Edited by Geoffrey Sunahara *et al.* (ISBN 0 471 986690)

Behaviour in Ecotoxicology
Edited by Giacomo Dell'Omo (ISBN 0 471 968528)

Predictive Ecotoxicology
Mark Crane (ISBN 0 471 984043)

Forecasting the Environmental Fate and Effects of Chemicals

Edited by

Philip S. Rainbow
Department of Zoology, The Natural History Museum, London, UK

Steve P. Hopkin
Department of Zoology, School of Animal and Microbial Sciences,
University of Reading, Reading, UK

Mark Crane
School of Biological Sciences, Royal Holloway,
University of London, UK

JOHN WILEY & SONS, LTD
Chichester • New York • Weinheim • Brisbane • Singapore • Toronto

Other Wiley Editorial Offices

John Wiley & Sons, Inc., 605 Third Avenue,
New York, NY 10158-0012, USA

WILEY-VCH Verlag GmbH, Pappelallee 3,
D-69469 Weinheim, Germany

John Wiley Australia, Ltd, 33 Park Road, Milton,
Queensland 4064, Australia

John Wiley & Sons (Asia) Pte Ltd, 2 Clementi Loop #02-01,
Jin Xing Distripark, Singapore 129809

John Wiley & Sons (Canada) Ltd, 22 Worcester Road,
Rexdale, Ontario M9W 1L1, Canada

Library of Congress Cataloging-in-Publication Data

Forecasting the environmental fate and effects of chemicals / edited by
Philip S. Rainbow, Steve P. Hopkin, Mark Crane,
    p. cm. -- (Ecological & environmental toxicology series)
  Includes bibliographical references and index.
  ISBN 0-471-49179-9 (alk. paper)
    1. Environmental toxicology--Forecasting. 2. Health risk assessment.
  I. Rainbow, P. S.   II. Hopkin, Stephen P.   III. Crane, Mark.   IV. Series.

RA1226 .F67 2001
615.9'02--dc21   00-051319

British Library Cataloguing in Publication Data

A catalogue record for this book is available from the British Library

ISBN   0 471 49179 9

Typeset in Garamond from the authors' disks by Dobbie Typesetting Ltd., Tavistock, Devon.
Printed and bound in Great Britain by Biddles Ltd., Guildford and King's Lynn.
This book is printed on acid-free paper responsibly manufactured from sustainable forestry,
in which at least two trees are planted for each one used for paper production.

Contents

List of contributors

Erik Baatrup
Institute of Biological Sciences, University of Aarhus, Building 135,
DK-8000 Aarhus C, Denmark

Mark Bayley
Institute of Biological Sciences, University of Aarhus, Building 135,
DK-8000 Aarhus C, Denmark

Patrick van Beelen
Laboratory for Ecotoxicology, RIVM — National Institute of Public Health and
the Environment, PO Box 1, 3720 BA Bilthoven, The Netherlands

Graham Blackmore
Department of Biology, Hong Kong University of Science and Technology, Clear
Water Bay, Hong Kong

Jack de Bruijn
National Institute of Public Health and the Environment (RIVM), PO Box 1, 3720
BA Bilthoven, The Netherlands

Mark Crane
School of Biological Sciences, Royal Holloway, University of London,
Egham, Surrey TW20 0EX, UK

Mark T. D. Cronin
School of Pharmacy and Chemistry, Liverpool John Moores University,
Byrom Street, Liverpool L3 3AF, UK

Raquel Duarte-Davidson
Environment Agency National Centre for Risk Analysis and Options Appraisal,
Steel House, 11 Tothill Street, London SW1H 9NF, UK

Steve Hopkin
Division of Zoology, School of Animal and Microbial Sciences, University of
Reading, PO Box 228, Reading RG6 6AJ, UK

Tjalling Jager
National Institute of Public Health and the Environment (RIVM), PO Box 1, 3720
BA Bilthoven, The Netherlands

R. W. P. M. Laane
National Institute for Coastal and Marine Management (RIKZ), PO Box 20907,
2500 EX, The Hague, The Netherlands

Michael C. Newman
Virginia Institute of Marine Science, College of William and Mary, Gloucester
Point, VA 23062, USA

Anthony O'Hagan
Department of Probability and Statistics, University of Sheffield, Sheffield
S3 7RH, UK

Paul Johnston
Greenpeace Research Laboratories, Department of Biological Sciences, Prince
of Wales Road, University of Exeter, Exeter EX4 4PS, UK

Simon J. T. Pollard
Environment Agency National Centre for Risk Analysis and Options Appraisal,
Steel House, 11 Tothill Street, London SW1H 9NF, UK

David I. de Pomerai
School of Biological Sciences, University of Nottingham, Nottingham NG7 2RD,
UK

Leo Posthuma
Laboratory for Ecotoxicology, RIVM — National Institute of Public Health and
the Environment, PO Box 1, 3720 BA Bilthoven, The Netherlands

Rowena S. Power
School of Biological Sciences, University of Nottingham, Nottingham NG7 2RD,
UK

Philip S. Rainbow
Department of Zoology, The Natural History Museum, Cromwell Road, London
SW7 5BD, UK

Morris H. Roberts Jr.
Virginia Institute of Marine Science, College of William and Mary, Gloucester
Point, VA 23062, USA

Michiel Rutgers
Laboratory for Ecotoxicology, RIVM — National Institute of Public Health and the Environment, PO Box 1, 3720 BA Bilthoven, The Netherlands

David Santillo
Greenpeace Research Laboratories, Department of Biological Sciences, Prince of Wales Road, University of Exeter, Exeter EX4 4PS, UK

Ton Schouten
Laboratory for Ecotoxicology, RIVM — National Institute of Public Health and the Environment, PO Box 1, 3720 BA Bilthoven, The Netherlands

T. Wayne Schultz
Waste Management Research and Education Institute, The University of Tennessee, PO Box 1071, Knoxville TN, 37901-1071, USA

Glendon D. Sinks
College and Veterinary Medicine, The University of Tennessee, PO Box 1071, Knoxville, TN 37901, USA

H. L. A. Sonneveldt
National Institute for Coastal and Marine Management (RIKZ), PO Box 20907, 2500 EX, The Hague, The Netherlands

Freddy F. Sørensen
Institute of Biological Sciences, University of Aarhus, Building 135, DK-8000 Aarhus C, Denmark

David J. Spurgeon
Institute of Terrestrial Ecology, Monks Wood, Abbots Ripton, Huntingdon, Cambs PE17 2LS, UK

Ruth Stringer
Greenpeace Research Laboratories, Department of Biological Sciences, Prince of Wales Road, University of Exeter, Exeter EX4 4PS, UK

Gunnar Toft
Institute of Biological Sciences, University of Aarhus, Building 135, DK-8000 Aarhus C, Denmark

M. Tonkes
Department of Water Pollution Control, Institute for Inland Water Management and Waste Water Treatment (RIZA), PO Box 17, 8200 AA Lelystad, The Netherlands

Foreword

Ecological & Environmental Toxicology is published by John Wiley & Sons and attempts to fill the gaps in the existing 'ecotoxicological' scientific literature. This latest book in the series *Forecasting the Environmental Fate and Effects of Chemicals* edited by Philip Rainbow (The Natural History Museum, London, UK), Steve Hopkin (Reading University, UK) and Mark Crane (University of London, UK), with 15 chapters by 31 international authors attempts to fulfil just such a role.

Their book considers the question of forecasting which is 'to predict, estimate or calculate beforehand; or a calculation or estimate of something in the future' This area is of great interest to ecotoxicologists whom attempt to predict effects at various ecological levels. It was no easy task for the editors to blend the varied contributions to this book as a wide range of issues are dealt with and in considerable detail. The editors have risen to the challenge of producing a book that is useful in its entirety but one that also enables readers to dip in to specific chapters.

We hope that this most recent contribution to the series will find favour with a wide range of readers due to the diversity of subject areas identified and succinctly reviewed. This book will be useful to scientists whom need to understand the status of current work and forecast solutions to ecotoxicological problems.

Jason M. Weeks
Sheila O'Hare
Barnett Rattner

Introduction

MARK CRANE[1], STEVE P. HOPKIN AND PHILIP S. RAINBOW[3]

[1]School of Biological Sciences, Royal Holloway University of London, UK
[2]Division of Biology, School of Animal and Microbial Sciences,
University of Reading, Reading, UK
[3]Department of Zoology, The Natural History Museum, London, UK

Most environmental toxicologists and chemists need to forecast the fate and effects of chemicals by using existing information to predict the future. This book, based on the output of the 9th Annual Meeting of the Society of Environmental Toxicology and Chemistry SETAC-Europe (UK Branch), 1998 is the first to address specifically the issue of forecasting in environmental toxicology and chemistry from a methodological, regulatory and practical perspective. It provides a unique and timely opportunity for active researchers and students to obtain in a single volume the 'state of the art' in an expanding field.

Crane sets the scene for the book in Chapter 2 in a wide discussion of the limits to forecasting in ecotoxicology. He emphasizes the importance of the need to know the toxic effects of chemicals **before** they are released into the environment, although quantifying the risks of new substances is extremely difficult. It is hard to achieve a balance between over-optimistic and over-pessimistic predictions. However, recommendations are made on how a compromise may be achieved between minimizing environmental damage without overly inhibiting economic progress.

In Chapter 3 O'Hagan introduces Bayesian statistics in the context of environmental toxicology. To Bayesian statisticians, the 'frequentist' techniques that most scientists currently use to assign probabilities are both philosophically and practically wrong. O'Hagan shows why this is the case, and describes how the Bayesian approach overcomes these problems to provide more comprehensible statements about the probability of future events. Newman and Roberts continue this theme of innovation in Chapter 4 and argue that periodic re-evaluation of paradigms and methods is essential as the science of ecotoxicology matures, and demands for improved accuracy intensify in ecological risk assessment. Current methods have emerged from a tradition biased towards point sources, single

Forecasting the Environmental Fate and Effects of Chemicals. Edited by Philip S. Rainbow, Steve P. Hopkin and Mark Crane.
© 2001 John Wiley & Sons Ltd

toxic agents, constant addition and effects on individuals, but forecasts now required for assessments must encompass complex exposure scenarios and effects on populations and assemblages. Newman and Roberts believe that forecasting can be improved by eschewing conventional concepts and techniques in favour of those more appropriate for present needs, and support this premise with five illustrations. Together, these examples show a need for more innovation, rather than more detailed applications of conventional tools for forecasting ecological risk.

In Chapter 5 Pollard and Duarte-Davidson describe the way in which the Environment Agency of England and Wales uses pollutant transport models to forecast the transport and fate of contaminants in the geosphere, and how these models fit within a risk assessment framework. Contaminated land models such as these face particular challenges in providing adequate representations of the real world, because of high system heterogeneity, multiple phases and analytical difficulties, all of which hinder model validation. Three case studies are presented, which give a flavour of the types of predictive models currently considered for use by environmental regulators within Europe. Jager and de Bruijn continue the European theme in Chapter 6, taking up one of the approaches described by Pollard and Duarte-Davidson. The European Commission's *Technical Guidance Document (TGD) on Risk Assessment of New and Existing Chemicals* was published in 1996 and is the currently accepted method for forecasting chemical risks to receptors in Europe. However, the simple deterministic approaches used to derive risk quotients have been criticized for a lack of environmental realism. Jager and de Bruijn provide a balanced account of the strengths and weaknesses of the *TGD* and make some useful and specific suggestions about how it can be improved.

The *TGD* deals with single substances, but complex mixtures of chemicals are often released into the environment as effluents from sewage treatment works and industrial processes. In Chapter 7 Tonkes examines whether the toxicity of these effluents can be predicted on the basis of toxicity data for known components. If this were the case, then regulation based upon chemical analysis would usually suffice and direct bioassays of effluents would be unnecessary. However, Tonkes found that in a survey of Dutch effluents the known toxicity of individual effluent components could only explain a small percentage of overall toxicity, so forecasts of toxicity based on chemical composition alone were of limited use. Mixture toxicity is also considered by Hopkin and Spurgeon in Chapter 8, this time in terrestrial systems. They consider the problem of determining which metal in a 'cocktail' of contamination near a smelter is responsible for toxic effects. Cadmium contamination is almost always found in association with zinc, but it is the cadmium that receives a bad press. Hopkin and Spurgeon show through the development of a simple relative toxicity factor, that the 'benign' and essential zinc can be the metal responsible for observed soil toxicity.

Several of the earlier chapters present ideas and data within a risk assessment framework that is increasingly accepted by environmental scientists and regulators. In Chapter 9 Santillo, Johnston and Stringer question the scientific and ethical basis of this risk assessment paradigm, contrasting it with a version of the precautionary principle. In Europe, the precautionary principle is enshrined in several laws, yet it seems at odds with other statutes which demand that risks be balanced against costs. This thought-provoking chapter is part of an ongoing debate about the level of chemical risk that we are prepared to tolerate in modern societies.

The remaining chapters in this volume describe a range of different approaches to forecasting the fate or behaviour of chemicals in the environment. The first two chapters describe approaches that use very simple test organisms. The ciliate *Tetrahymena pyriformis* has proved to be an excellent model for studying the toxicity of chemicals in fresh waters. Cronin, Sinks and Schultz tested the effects of 140 aliphatic compounds on the ciliate and, in Chapter 10, show that quantitative structure–activity relationships (QSARs) can be used, in almost all cases, to predict toxicity from knowledge of physico-chemical structure alone. The five exceptions to the general rule were small reactive molecules (e.g. acrolein). In Chapter 11 Power and de Pomerai describe an approach to soil biomonitoring with a transgenic stress-inducible nematode (the now famous and genomically defined *Caenorhabditis elegans*). The heat-shock or stress response represents a short-term, cellular response to protein-damaging events that may forecast effects at higher levels of organization. The induction of heat-shock proteins has traditionally been studied using immunodetection techniques, but this approach is time-consuming and costly. The study described in this chapter assesses a transgenic strain of the soil nematode *C. elegans*, carrying a stress-inducible β-galactosidase reporter, for use in toxicity testing of soils. Reporter enzyme activity provides a quantitative measure of the induction of the heat-shock response, and the presence of cadmium or mercury, but not of copper or zinc alone, in the soil significantly induced this reporter response. A rapid soil-toxicity assay could well be developed based on this transgenic strain of *C. elegans*.

Baatrup and colleagues at the University of Aarhus have pioneered the study of changes in animal behaviour as an ecotoxicological indicator of stress due to exposure to a wide range of substances. In Chapter 12 Baatrup, Bayley, Sorensen and Toft discuss the extent to which changes in animal behaviour can predict effects at the population level. Developments in computing have allowed the movement of animals to be quantified for extensive periods of time using automated video tracking. For example, exposure of springtails to dimethoate disrupts recognition of conspecifics and aggregation behaviour, disorders which are likely to impair the growth, survival and reproductive success of populations in the field. The locomotory behaviour of woodlice differs markedly between populations collected at clean sites and those from a region contaminated by fallout after a fire at an industrial plastics factory. Recent work has also shown that the sexual behaviour of guppies is inhibited at very low concentrations of endocrine disruptors.

Moving from populations to assemblages of organisms, in Chapter 13 Posthuma, Schouten, van Beelen and Rutgers have tested whether the 'added risk' approach is appropriate for deriving benchmark concentrations for soils. The method defines hazardous concentrations (HCs) on the basis of a frequency distribution of sensitivities of organisms. Responses to zinc exposure in nematodes and micro-organisms were measured. The HC5 concentration (hazardous for 5% of species) was found to be a meaningful measure of small or negligible risk to the soil community, whereas the HC50 (hazardous for 50% of species) indicated areas in need of remediation.

The final two chapters show how both chemical analytical and biological approaches can be used to forecast the distribution of contaminants in marine systems. Sonneveldt and Laane in Chapter 14 faced the challenge of forecasting sediment quality in European coastal waters using a sediment-exchange model. The annual average particulate loads of Cd, Cu, Pb, Zn and PCBs in surface sediments of the Dutch coastal zone decreased in the 1980s and stabilized in the 1990s, these decreases being attributable mainly to decreases of loads from rivers and the dumping of dredged materials. The authors show that the trends in the observed concentrations of several compounds in the $<63\,\mu m$ fraction of surface sediments of the Dutch coastal zone could be reproduced quite well by a relatively simple sediment-exchange model for the period 1981–96. Their model is based on a sediment balance for the coastal zone, the mixing depth of the active top layer and a quantification of the annual average particulate loads to the area. The delay between the annual average load reduction and the simulated concentration is about two to three years. The results show that physical exchange of mud particles between the water column and the active bottom layer may be a dominant fate process for compounds in the dynamic sandy sediments of the Dutch coastal zone. The authors go on to use the model to predict the development of the concentrations of metals and organic compounds in the surface sediments of the Dutch coastal zone, with changing loads from different sources such as dumping of dredged materials.

In the final chapter, Blackmore and Rainbow have used trace metal biomonitoring techniques to define trends in metal bioavailabilities in a Hong Kong coastal habitat as a means of predicting the future status of local metal pollution. Bivalve molluscs and barnacles have proved to be useful tools for monitoring geographical and temporal patterns in trace metal bioavailability in coastal waters throughout the world. Samples of one mussel and two barnacle species were collected from Tolo Harbour and reference sites during April 1998, and existing spatial patterns of bioavailability inferred from accumulated metal body burdens. These data have then been compared to those of similar studies from 1986, 1989 and 1996 in order to discern long-term changes in metal bioavailabilities to these biomonitors over the previous 12 years. After a local increase in Tolo Harbour in the late 1980s, there has been a general decline in trace metal availabilities in the 1990s, correlating with the migration of local industry from central districts of Hong Kong, firstly to the New Territories including the Tolo Harbour region, and then from Hong Kong into southern China. Unless there is a further industrial development along the shores of Tolo Harbour, trace metal bioavailabilities in this enclosed habitat should continue to fall until stabilized.

The authors of each chapter include several examples which emphasize the importance of being able to predict the effects of chemicals released into the environment. However, they also highlight the difficulties. Considerable progress has been made in recent years, but it is clear that there is still some way to go. It is unlikely that predictions will ever be 100% accurate, but we are certainly getting closer to this goal.

2

Limits to Forecasting in Environmental Toxicology and Chemistry

MARK CRANE

School of Biological Sciences, Royal Holloway University of London, UK

'Forecasters in each field encounter the same challenges and make the same mistakes in their attempts to predict the future.'

William Sherden (1998) *The Fortune Sellers*

2.1 INTRODUCTION

None of us can see into the future, but most of us must try. We plan our future finances, our future careers and our future happiness on the basis of many assumptions. Some of these assumptions may be little more than guesses about what the future might hold. Others might be based on more solid foundations. If it were possible to distinguish pure guesswork from more reliable approaches to forecasting we could have more faith in the future, and this is as true for forecasting within scientific disciplines as it is for forecasting in our personal lives.

In environmental toxicology and chemistry we need to forecast what new chemical products may be released into the environment. We then need to forecast how they will behave and what their eventual fate might be in a host of different environmental compartments. Finally, and perhaps most importantly, we need to forecast any adverse biological effects that these chemicals may have along their pathways.

Not all environmental toxicologists and chemists are explicitly involved in forecasting the future. Some deal with historical contamination and appear to ask questions about the distribution and effects of toxicants in the present. However, even statements about the present are, of necessity, heavily tinged by expectations of the future. To say that a piece of land is contaminated with metals *now* is to imply that without remediation it will remain so into the future. To describe a chemical as toxic implies that it has always and will always be toxic. Hence the future is always with us.

In this chapter, what makes a good forecast is considered, as are some of the tools that can be used to forecast future events. Then the factors that shape human belief in forecasts are discussed and it is shown why these may lead to overly optimistic or pessimistic conclusions. Finally, some of the factors that limit our ability to see into the future are reviewed and how these should

Forecasting the Environmental Fate and Effects of Chemicals. Edited by Philip S. Rainbow, Steve P. Hopkin and Mark Crane.
© 2001 John Wiley & Sons Ltd

influence our selection of forecasts in environmental toxicology and chemistry is considered. In this chapter the ideas and terminology of others, particularly those in Nicholas Rescher's 1998 book *Predicting the Future: An Introduction to the Theory of Forecasting* have been heavily drawn upon. He claims that this is the first general theory of forecasting since Cicero wrote *De Divinatione* in the first century BC. The author is also indebted to Anthony Sherden, whose 1998 book *The Fortune Sellers* provided useful insights into several aspects of forecasting.

2.2 WHAT IS A GOOD FORECAST?

Over 2000 years ago Cicero drew a distinction between forecasts that have predictive value and those that simply serve a political purpose by providing expert approval for a particular course of action. Although it would be naive to assume that decision makers today do not still crave the shelter that is afforded by expert approval, forecasts that are not predictive of the future clearly have little value. Members of the last British Government had cause to regret incorrect forecasts of the impact of 'mad cow disease' made by civil servants. At the time they were made these forecasts served the Government's interests well by supporting a cheap 'do nothing' option, but the political fallout that resulted later when the full extent of the problem became apparent far outweighed the short-term financial and political gain.

A good forecast according to Rescher (1998) consists of both a good question and a good answer. **Questions** about the future should be important, interesting, resolvable and difficult. An **important** question is one where much could be gained or lost, depending on the answer. The manufacturers of a new pesticide could stand to lose large sums of money if their product is not approved for use, and the environment could suffer considerable damage if the product is approved inappropriately. Most people would probably feel that an accurate forecast of the effects of this pesticide would be important.

However, some important questions may be completely uninteresting to 'most people'. Do the public really care (would they notice?) if this pesticide reduced energy flows through a woodland by 10%? It is in this interplay between importance and interest that a dialectical process should, and often does, take place between the scientific and the wider community. Scientists need to communicate to the public what they have discovered to be important and need to pay due regard to what the public themselves believe is important and interesting.

Questions about the future should be **resolvable**, which seems obvious until one discovers that questions such as, 'what is the effect of pesticide x on ecosystem health' are rather common in environmental toxicology. The concept of ecosystem health has been debated elsewhere (e.g. Crane and Newman 1996). It is sufficient here simply to mention that there are doubts whether terms such as 'ecosystem health' can have any useful meaning, or identify a tangible phenomenon. If this is so, then resolving questions about concepts such as ecosystem health is probably impossible.

Less obvious is that good forecasting questions should be **difficult**. Easy questions are likely to be of little importance or interest. Will new toxic chemicals be manufactured next year? Will there be any pollution incidents over the next decade? Will new pollution remediation technologies be developed in the next century? All of these questions can be answered in the affirmative with some security, but such risk-free forecasts do not help us much. **Which** new chemicals will prove to be toxic next year? **Where** will pollution incidents occur in the next decade? **What** new pollution remediation technologies will be developed in the next century? These are more difficult, more risky and much more useful forecasting questions.

Good **answers** to questions about the future should be correct, accurate, relevant and detailed. Since it may be difficult to determine whether a forecast is correct for quite a long time to come (e.g. 'pollution incidents will decrease every year for the next century') it should at least be **credible** at the time it is made, either because the expert who made the forecast has a good track record of forecasting or because there is plenty of evidence available to make an inference. If a 'correct' forecast cannot be made it should at least be reasonably **accurate**. For example, if an Environmental Quality Standard of $10 \, mg \, l^{-1}$ of substance x is forecast to be safe, then future determination that the safe level is in fact $8 \, mg \, l^{-1}$ will show the forecast to be incorrect, but reasonably accurate. If the safe level of x actually turns out to be $0.008 \, mg \, l^{-1}$ then the forecast will have been both incorrect and highly inaccurate.

That answers should be **relevant** to the question that was posed might again seem obvious, but experts can often disagree over even apparently fundamental concepts and tools in their discipline. For example, there remains considerable disagreement among environmental toxicologists over whether the results of single species tests in the laboratory are relevant to questions about the impact of chemicals on species assemblages in nature (Crane 1997). This is of concern because single species toxicity tests remain the most common way of answering the forecasting question, 'what will be the impact of chemical x on natural assemblages?'

Finally, **detailed** answers are superior to vague ones because they are both more useful and easier to test. A forecast that particular concentrations of an effluent will be toxic to insect larvae in a river is more useful than forecasts that the effluent might be harmful to biota in general.

Unfortunately, despite such common-sense rules for defining a good forecast, humans often follow less sensible approaches, as discussed in the next section.

2.3 WHAT SHAPES BELIEF IN A FORECAST?

Humans have a variety of reasons for believing in particular forecasts, many of them based on rather eccentric worldviews. Our idea of what might happen in the future is often influenced by the situation in which we currently find ourselves, or by our most recent experiences. This **situational bias** might, for example, lead toxicologists and chemists working in industry to underestimate the potential hazard of a chemical, because they assume that the users of chemicals will always follow the instructions on labels. On the other hand, someone working for an environmental lobby group may regard all industrial discharges with alarm, and overestimate the dangers posed by them, because so much of their time is spent in investigating industrial pollution incidents. Situational bias can also result from, or perhaps is a cause of, wishful or fearful thinking. Some of us are temperamentally inclined to hope for the best, while some fear the worst, and this will colour our perception of forecasts. Many humans are also innately conservative, which can lead to an exaggeration of the stability and durability of the current order (Rescher 1998). For example, it is hard for some people to worry much about the effects of global warming, because to them weather patterns do not seem that different from the way they have always been.

The **immanency exaggeration** occurs when people exaggerate the immanency or extent of a future event (Rescher 1998). Promising breakthroughs in science, medicine and technology are often reported to be 'just around the corner'. While self-interest and the desire to secure additional funding undoubtedly play a part, there is often a genuine belief in these claims on the part of those making them.

Finally, **probability misjudgement** can occur because most humans have rather a poor grasp of probability concepts. Probability misjudgement occurs when incorrect evaluations of single probabilities or incorrect combinations of probabilities lead either to an overestimation that unlikely events will occur, or an underestimation that events with a small probability will happen (Rescher 1998). For example, a pessimist might take the view after several years without a major oil spillage at sea that the 'laws of probability' predict that an accident is due. This view is nothing to do with probability. On the other hand, an environmental manager may take the view that since the probability of toxic effects caused by chemical x is low and the probability of environmental exposure is also low then there is no chance of a pollution event caused by the chemical. This is a naive approach to combining probabilities. The frequency with which humans commit these fallacies may be because under certain circumstances such views are correct. A belief in the idea that an event is about due is true for phenomena where exhaustion (e.g. of a food supply) or saturation (e.g. of existing farmland) may occur. Also, phenomena that are either true or false, rather than probabilities, do combine to produce correct forecasts by simple combination.

Those involved in the sciences of environmental toxicology and chemistry should obviously resist the fallacies listed above if accurate forecasts are required. But scientists are only human, so what can they do? Personal awareness of potential fallacies is a first step, the creation of scientific teams in which even the most junior members can question the most senior would be a useful second step, and the peer review of all outputs would be an effective final

Table 2.1 Approaches to forecasting (after Rescher 1998)

Forecasting approach	Linking mechanism	Method of linkage	Forecasting condition
Intuitive			
Judgemental estimation	Expert panel	Informed judgement	Learnable (orderly)
Inferential			
Trend projection	Current trends	Projection of current trends	Trend uniformity
Curve fitting	Geometric patterns	Line of best fit through available data	Stable temporal patterns
Analogy	Grouping of comparable phenomena	Incorporate into analogous situation	Actual (rather than apparent) analogy
Indicator coordination	Correlation	Statistical evidence of a correlation	Stable correlations
Modelling	Formal models (physical or mathematical)	Analogy between model processes and real world processes	Fixed structural mode of operating
Law derivation	Accepted laws (deterministic or statistical)	Inference from accepted natural laws	Stably lawful (regular)

step. Many scientists already operate according to these criteria. Unfortunately, self-interest and delusion, traditional management structures, particularly in academia, and the lack of expert peer review for many 'grey literature' reports commissioned by government agencies can undermine all three steps.

Assuming that there is a will to forecast the future in as accurate and unbiased a way as possible, what tools are available?

2.4 WHAT FORECASTING TOOLS ARE AVAILABLE?

Forecasts can be made according either to the intuition of experts or to transparent inferential decision rules. Either way, rational forecasting depends upon three conditions (Rescher 1998):

1. **Data availability** (there must be data upon which to base a forecast).
2. **Pattern discernibility** (there must be a discernible pattern in the data).
3. **Pattern stability** (the pattern must be stable, rather than random or chaotic).

Table 2.1 lists rational approaches to forecasting that have been used successfully in the past. Of course, more than one approach can be used to forecast a future event, and if several different forecasting approaches each yield a similar forecast then we can have more faith in the correctness of the forecast.

2.4.1 EXPERT JUDGEMENT

Experts may have intuitive judgements about the future that cannot be properly formalized, particularly if their domain of expertise provides learnable, orderly lessons. The use of expert panels is widespread in environmental toxicology and chemistry. The Society of Environmental Toxicology and Chemistry, organisers of the meeting that generated this book, frequently runs workshops in which experts are asked to make forecasts about the future fate, behaviour and effects of potential toxicants (e.g. Chapman *et al.* 1996, de Fur *et al.* 1999, Stebbing *et al.* 1993). However, reliance upon expert judgement can lead to problems. Who selects the experts, and what if they disagree? According to Rescher (1998), good forecasters should be selected for their strengths in several areas:

1. **Systematic consistency** (either absolute reliability in that they forecast correct answers, or comparative reliability in that their forecasts are better than those made by others).
2. **Versatility and range** (ability to forecast over a range of conditions).
3. **Daring** (ability to deal with difficult issues).
4. **Perceptiveness** (ability to provide detailed and definite forecasts).
5. **Foresight** (ability to see further into the future than others).
6. **Consistency** (uniform performance over time).
7. **Self-criticism** (ability to indicate greater confidence when more certain of a forecast).
8. **Knowledgeability** (understanding of non-forecasting issues in their domain of expertise).
9. **Coherence** (forecasts are compatible with one another).

One problem with expert groups is that group interactions and peer conformity can lead to rather dull and conservative forecasts that represent an 'average' view. The best that can be said for averaging is that, '. . . the average will always be closer to the actual outcome than the **worst** (my bold) of its component individual predictions' (Rescher 1998). Other than this, it will usually be unclear whether an average forecast is better or worse than each individual forecast. For example, five experts might forecast that $5 \ \text{mg} \ l^{-1}$ is a safe concentration for chemical x, and four might forecast that $50 \ \text{mg} \ l^{-1}$ is the safe concentration. The average forecast would therefore be a safe concentration of $25 \ \text{mg} \ l^{-1}$. Once the chemical is released into the environment the $5 \ \text{mg} \ l^{-1}$ estimate may emerge as correct, a value forecast by most of the experts, but five times lower than the average forecast. On the other hand, the average would clearly have been superior to all individual forecasts if the true safe concentration of chemical x lay in the region of $20\text{--}30 \ \text{mg} \ l^{-1}$.

A more formal approach for obtaining expert consensus forecasts is to use the Delphi method. This involves the elicitation of information from experts, '. . . punctuated by feedback stages in which earlier responses are conveyed to the group in a condensed, statistically summarised form' (Rescher 1998). The approach is designed to release group members from the pressures of group interaction and peer conformity by making responses anonymous. The responses of participants can also be weighted for their level of expertise. However, the danger remains that the consensus estimate will follow a line of least resistance to a dull and incorrect conclusion, especially when the experts are conservative. Woudenberg (1991) found no evidence that the Delphi method is any more accurate than other methods for obtaining expert judgement and that consensus is still achieved by conformity to peer group pressure.

Instead of using experts throughout the forecasting process, **expert systems** can use just the methodologies elicited from experts to produce forecasts on the basis of these methodological rules. Unfortunately, the rules themselves may still be tainted by the prejudices of the human experts who constructed the methodologies.

It seems then that sole reliance on expert judgement for forecasting should be used only as a last resort if other approaches are unavailable. After all, 'The history of science amply illustrates that consensus in error is no less common than consensus in truth' (Rescher 1998).

2.4.2 SIMPLE INFERENTIAL METHODS

More formal inferential methods are available if we choose not to rely upon expert judgement alone. The simplest of these is **trend projection** when the phenomenon is expected to have a uniform trend. For example, the best approach to take when forecasting the future direction of science and technology, such as new remediation technologies for contaminated land, is **pipeline analysis** in which research and developments that are currently under way ('in the pipeline') are projected forward to make a forecast. However, this approach cannot tell us anything useful about long-term trends not yet in the pipeline, it tends to be conservative and does not reflect 'non-consensus' developments with potentially high impact. It also remains very difficult to estimate the speed at which technologies in the pipeline will emerge.

Phenomena that are completely anarchic and occur by chance cannot be forecast, by definition. However, if events change at random over time (rather than chaotically) and we know nothing about them, then the best forecast possible is that in the future there will be no change from the current situation, i.e. trend projection (Pindyck and Rubinfeld 1976).

Analogy can be used when the phenomenon is expected to maintain an actual, rather than just an apparent, analogy with another phenomenon. For example, the production, fate and behaviour of chemical y could be compared to a very similar chemical x manufactured 10 years earlier. The problem with analogy is in identifying the impact on forecasts that may be caused by **differences** between the analogized systems, rather than just concentrating on the areas of similarity. What if production of chemical y is much greater than forecast because of improved marketing? What if a new use is found for it that changes exposure patterns?

Other simple inferential methods are curve fitting to time series and cyclical analysis when the phenomenon is expected to follow stable temporal patterns. Unfortunately, it can be difficult to identify the current stage in a cycle. Because of the problems associated with simple inference, more sophisticated approaches have been developed.

2.4.3   SOPHISTICATED INFERENTIAL METHODS

More sophisticated forecasts may be based upon forecasting indicators, natural laws or modelling. **Forecasting indicators** use correlations between phenomena to forecast one from another, so long as these correlations are expected to be stable. Environmental toxicologists do this frequently when they use the results from single species toxicity tests to forecast effects on populations and assemblages of species. Indicators (and some models) can have a known mechanistic link or be 'black boxes' in which we know only that there is a proven track record of successful correlations between the indicator and the indicated. Clearly, it is best to have an understanding of the link between the two so that a correlation is not assumed under inappropriate conditions. Complete mechanistic understanding is not essential, and may not be possible if a forecast is required rapidly on the basis of only limited information. However, in most cases it is likely that explanations of the natural world and good forecasts are symbiotic (Rescher 1998). This is because good explanatory theories are usually established as such because of past predictive successes and because they are embedded in a wider framework of explanatory theories. The forecasts we accept also usually need some sort of theoretical underpinning to make them more credible than other alternatives.

Physical and mathematical **models** are used in environmental toxicology and chemistry and may be the only way to forecast some future events. They are appropriate if the phenomenon that is modelled has a fixed structural mode of operating. However, great caution needs to be applied since the enthusiasm of modellers can cause them to concentrate more on the elegance of their model than on its application to the real world. Complex models have been found lacking in several fields. In economics the use of complex models does not lead to better predictions than the use of subjective judgement (McNees and Ries 1983), and Sherden (1998) quotes several authorities who suggest that in human demography complex models have also proved to be of limited use:

- '[there is] . . . no evidence that complex and/or sophisticated techniques produce more accurate or less biased forecasts than simple naive techniques . . . Expertise has been shown to add little to forecast accuracy' (Ahlburg and Land 1992).
- '. . . simple projection techniques are more accurate than more complex techniques' (Stoto 1983).
- '[complex models] . . . have not led to greater accuracy in forecasting total population than can be achieved with simple naive techniques' (Smith and Sincich 1992).

If these criticisms are valid for human populations then we should question whether projections from complex models could possibly be better for phenomena such as wildlife populations, whose births and deaths are usually far less easy to measure.

Environmental toxicologists have historically focused on survival and birth rates in their tests. However, the demographics of any population depend upon these two parameters **plus** migration rates: either immigration into an area or emigration from it. We may question whether toxic effects on migration rates could ever be forecast reliably for wildlife populations. This view is supported by the difficulties encountered in forecasting even human migration rates (Sherden 1998).

Leaving the specific case of demographic models to one side, the generic problems with mathematical models usually hinge on insufficient data to support model assumptions, and analytical complications caused by the complexity of the real world. Both the economist Milton Friedman (1953) and the ecologist Robert Henry Peters (1991) have argued strongly that it is predictive ability and reliability that is the only hallmark of a good model. Realism is immaterial. However, as mentioned earlier, the reflection of at least some knowledge of mechanism in the models we use will increase our confidence in model outputs.

Finally, the best approach to forecasting is usually one based on inference from formalized **scientific laws**. However, as we shall see later, even forecasts based on scientific laws can prove incorrect if chance or chaos intrude, or in domains '. . . whose phenomenology is so complex as to put the operative laws outside the range of our cognitive vision' (Rescher 1998). For example, community ecology arguably occupies such a domain, which makes accurate forecasts of assemblage structure highly problematic.

2.4.4 PROBABILISTIC FORECASTS

Ideally our forecasts, whatever their basis, will state categorically that a certain event will occur at a certain time in the future. However, where certainty is lacking we may often have to rely on probabilitic forecasts that a particular outcome is likely. These types of forecasts can be problematic. What are we to make of a forecast 50% chance that it will rain? We might just as well flip a coin if we need to make a decision based on a 50:50 chance. It is also impossible to falsify forecasts like these. Similarly, how can an environmental manager use a forecast such as 'there is a 10% risk of toxicity to 40% of species in a biological assemblage 20% of the time in the river below this effluent discharge?'

Probabilitic forecasts are of use not as absolute guides to decision making, but as relative guides. Our decision to carry an umbrella may be influenced by a forecast of 50% rain if the daily forecasts for the previous month have all been zero chance of rain (and have all proved correct). Similarly, we may be alarmed or reassured by a 50% chance of fish mortality if previous forecasts of an effluent's toxicity have been nearer 0% or 100%. The point here is that probabilitic risk assessments provide us with information on relative dangers, rather than certain forecasts of harm.

So, we have some tools and approaches, albeit none that is perfect, with which we can forecast the future. But are some phenomena more amenable to forecasting, and are there any general limits to what, and how far into the future, we can forecast?

2.5 ARE THERE LIMITS TO FORECASTING?

2.5.1 THE PROBLEM OF INDUCTION

Scientists and philosophers have argued for many years over the relative merits of inductive and deductive logic. Inductive approaches concentrate on the gathering of data in the hope that useful generalizations will emerge. Deductive approaches usually emphasize the testing of a hypothesis by gathering critical data that could falsify it. Crane and Newman (1996) discuss the use of inductive and deductive knowledge in environmental toxicology, and suggest that both approaches have value, although induction will clearly be the most important method when forecasting, because of the need to generalize with only the data in hand. Deductive methods may be useful later to test whether a forecast was correct.

Unfortunately, a big problem with inductive approaches is that induction is itself required to justify the approach. Why do we believe inductively that the sun will rise at dawn tomorrow? Because the sun has always risen at dawn. Why should we use induction? Because inductive approaches have often worked in the past. In other words, the **problem of induction** is that it relies on a circular argument. This problem of induction is a real logical obstacle to forecasting and is the sort of difficulty that makes many philosophers uncomfortable. Karl Popper believed that induction had no, or at best a limited place in science and that the success of a theory in the past was no guide to its success in future (Crane and Newman 1996). Rescher (1998) takes issue with Popper's contention that 'the degree of corroboration of a theory merely [serves] as a critical report on the quality of past performance: it could not be used to predict future performance'. He believes that this is an absurd position to take, '. . . **of course** past performance is a predictive indicator. (What could possibly serve better?) What past performance does not enable one to do is predict with failproof accuracy.' In his view it is irrational to ask for the impossible: inductive approaches to forecasting offer a better hope for success than any other known alternatives. Hence, if we want to forecast the future, we just have to put up with the problem of induction.

A more tractable disagreement is over the nature of phenomena in the real world. Over the centuries philosophers have disagreed on the extent to which the world's events are determined or random. Greek Stoics took the view that the world was completely deterministic, a belief shared later by Laplace and, apparently, by many modern-day scientists. According to this view if we could only discover the laws underlying natural processes we should be able to forecast future events with certainty. In the next sections factors other than the problem of induction that limit forecasts are investigated, in an attempt to show that although more knowledge can be attained through science, some of our knowledge reveals the impossibility of completely accurate forecasts for some phenomena. Unfortunately, some of these least forecastable phenomena are of considerable interest to environmental toxicologists in particular.

2.5.2 EFFECTS OF INADEQUATE INFORMATION

Forecasts can be wrong either because the information used was incorrect or because it was incomplete. Environmental managers know very well that decisions usually need to be made in the absence of perfect knowledge. Perhaps the information could theoretically be collected, but this is practically impossible within the time and other resources available. For example, in the United States the Toxic Substances Control Act requires the Environmental Protection Agency to make a regulatory decision about new industrial chemicals with only limited data and within a very short time span (Zeeman 1995).

Alternatively, there may be a mismatch between the parameters needed for theory-based forecasts and the parameters that can actually be observed and measured. It is then necessary to use estimates, with all their attendant errors.

Both **analysis over-determination** and **under-determination** can result from inadequate data (Rescher 1998). With analysis over-determination there is a surfeit of plausible forecasts that can be made on the basis of the data, all pointing in different directions. The results from a mesocosm test with an insecticide might suggest that most arthropods are killed, but that the majority of the populations recovered within a few weeks. I suspect that a scientist working for an environmental lobby group would use these data to forecast a rather different scenario in nature to that forecast by a scientist working for an agrochemical company.

With analysis under-determination there are simply no useful data to support any particular forecast among a group of alternatives. Are freshwater invertebrates being impacted by endocrine disrupting chemicals? We currently know too little about the endocrinology or environmental exposure of invertebrates to be sure (de Fur et al. 1999).

2.5.3 EFFECTS OF ANARCHY

Some events occur purely by chance. Whether the next flip of a coin will produce heads or tails is something that cannot be forecast. Where and when will the next collision occur between an oil tanker and another vessel? Will the wind blow the oil towards the shore or away from it? These types of questions are intrinsically unanswerable. All one can do is prepare for the worst case, without knowing whether it will ever occur.

2.5.4 EFFECTS OF CHAOS

Chaos is the term given to erratic but bounded oscillations in any phenomenon. The concept has been popularized by authors such as Gleick (1987), but is still often misunderstood. Chaos is not the same as anarchy. Chaos occurs in non-linear systems that are sensitive to small changes in initial conditions. Although the behaviour of these systems is deterministic it leads to unpredictable changes over time. May (1974) was one of the first to identify the potential for chaos in populations of organisms. Since populations are one of the main foci for environmental toxicologists, this has two important implications for the forecasting of toxic effects (Newman 1995). First, chaotic systems should be forecastable over short periods, but not over long periods. Second, improved forecasting accuracy for non-linear systems depends upon our ability to improve the measurement of initial conditions, which is not at all easy. This is particularly so if an ecological equivalent of the Heisenberg uncertainty principle applies to our measurements and the act of measuring influences the course of events.

Sherden (1998) draws the following conclusions about the reliability of human population forecasts:

1. Forecasts of future cohort populations that already exist are quite accurate.
2. Forecasts of population counts are more reliable than forecasts of population growth rates.
3. Periods of steady growth allow greater forecasting accuracy than periods of change.
4. Forecast accuracy is worse when populations are growing rapidly.
5. Forecasts of populations within large geographical areas are more accurate than those for smaller areas, probably because differences in migration rates do not need to be accounted for.

These comments on the reliability of human population forecasts must apply with even greater force to wildlife populations in which mortalities, births and migrations are usually more subject to change and are more difficult to measure. Caswell (1996) contends that there is a distinction between a population **projection**, which is what demographers like to think they do, and a population **forecast**. A projection is simply a demographic scenario that follows logically from the assumptions of the model that is used. However, if a projection is not a forecast then is it of any use to those that wish to know the future? Of course, there is nothing wrong with scenario construction as such; it is undoubtedly an interesting and instructive venture, and keeps many scientists and even more tax dollars gainfully employed. However, the fact remains that scenarios are only a matter of surveying possible courses of future development. 'They are imaginative speculations about what **might** happen and not informative specifications attempting to preindicate what **will** happen' (Rescher 1998). In fact, what tends to happen is that 'a demographer makes a projection and his reader uses it as a forecast' (Keyfitz 1981).

2.5.5 EFFECTS OF COMPLEXITY

Complexity is a term increasingly used to describe the study of order emerging from systems in which complex interactions are guided by an overarching principle (Pagels 1988). Evolution is a much-cited example. Individuals of different species have interacted with each other over millions of years to form complex assemblages 'guided' by the principle of natural selection. Horgan (1996) draws little distinction between researchers of chaos and those who study complexity, describing both as practitioners of 'chaoplexity'. However, there are some useful differences between the study of chaos and complexity.

According to Sherden (1998), complex systems are unpredictable because they

- have no, or rather weak, laws governing their behaviour at lower levels of organization
- cannot be reduced into their component parts because they arise from the interactions between the parts
- are highly interconnected by positive and negative feedback loops
- alternate unpredictably between periods of order and turmoil
- adapt to their environments and evolve new strategies that invalidate theories based upon their previous behaviour
- have no fixed cycles and their histories do not repeat themselves.

Sherden (1998) uses economic examples to illustrate complexity, but it is interesting to note that the points listed above could apply to most biological species assemblages. In contrast to populations, which may behave chaotically, assemblages or communities of species are complex systems. For a start, there is usually no way of forecasting when one organism will interact with another, and if the concept of 'community' has any meaning at all then it must refer to interactions between individuals and species. There are many positive and negative feedback loops in the dynamic food webs drawn by community ecologists (Lawton 1989), and there is ample evidence that many assemblages alternate between different stable states (May 1989). Individual organisms are able to adapt to their environments, and the view has been advanced that history does not repeat itself when assemblages of organisms are analysed over time (Matthews et al. 1996).

'Complex organic systems such as species or ecologies or societies are adaptive rather than deterministic in that the rules change in the light of the consequences of the behaviour they produce' (Caulkin 1995). This makes the structure or functioning of complex systems impossible to forecast accurately, although forecasts of general trends might be possible.

2.5.6 EFFECTS OF FACTOR EXFOLIATION

Factor exfoliation occurs when causality depends upon several factors that are difficult to determine and which, in turn, depend upon other factors that are difficult to determine.

We . . . now find that [our forecast] has exfoliated into a number of others each of which in turn exfoliates into various others each of which has its own problems and difficulties, and many of which belong to domains very remote from that of the first. Issues of this factor-exfoliating sort can readily prove to be predictively intractable because the outcome becomes veiled in the fog of a complexity into which we have — and can obtain — little or no secure insight' (Rescher 1998).

A common question in environmental toxicology is whether 'links' can be determined between effects at different levels of biological organization. This is an example of factor exfoliation. Can we forecast the effect that a change in bioaccumulation or biomarker activity might have on population parameters, assemblage structure and, ultimately, on ecosystem function? The opportunities for anarchy, chaos and complexity to frustrate such an attempt appear almost endless.

2.5.7   EFFECTS OF SCALE

Different scales in time and space may also frustrate accurate forecasts, although problems can occur at all scales. 'At every level of scale in cosmic organisation — the atomic, the biological, the cosmic — we seem to encounter a mixture of stability and instability, of order or anarchy, of regularity and fluctuation, of predictability and unpredictability' (Rescher 1998). If phenomena are at the end of a hierarchical chain then they may be particularly difficult to forecast, as mentioned above.

Biological scales such as generation time and body size should be taken into account when forecasting. For example, factors affecting relatively long-lived and wide-ranging fish species may be more appropriately forecast at the watershed scale, while factors affecting short-lived and relatively sedentary macroinvertebrate species may be best forecast at the stream segment scale.

2.5.8   EFFECTS OF CHANGE AND INNOVATION

Human innovation is the most likely future change that may prove our forecasts to be incorrect. Changes in the use or disposal patterns of toxic chemicals. Changes in methods of toxicological assessment or remediation. Changes in political and economic pressures. All of these may impact upon the veracity of forecasts about the fate and effects of toxic chemicals. Serendipity, the accidental discovery of useful phenomena, could play a part in unhinging our forecasts. Non-human organisms may also change. Among higher animals learnt behaviour may cause them to avoid areas of pollution. Over longer timescales natural selection may change tolerance or resistance to contaminants.

We cannot even be sure about future science for we do not know the questions that future science will ask. Could environmental toxicologists even 10 years ago forecast the current high level of interest in so-called endocrine disrupting chemicals? Do we know that this apparently important problem will still be considered so in another 10 years? We do not know whether future science will be able to answer questions that we currently believe to be intractable and we also cannot be sure that future science will not support ideas that we currently reject.

However, this does **not** mean that we should all give up and go home. There is an important distinction between those forecasting questions that we cannot answer simply because we are ignorant and those that we cannot answer because chance, chaos or complexity intervene. In the case of ignorance we would be wrong to assume that just because science cannot currently provide good forecasts it will never be able to do so. In the case of chance, chaos and complexity we should be more secure in our belief that a forecast is possible only within certain limits. This is because we have succeeded in **explaining** the impossibility of an accurate forecast. If a population of organisms or the distribution of a chemical fluctuates chaotically then it is **by definition** impossible to forecast the size of the population or the concentration of the chemical beyond a certain point in time. If one is asked to provide such a forecast then the forecasting question is simply **illegitimate** (Rescher 1998).

2.6 CONCLUSIONS

Different predictive approaches are superior on a case-by-case basis, but there are some general principles that the pathologically risk averse can follow:

1. **Keep it dull**. Although the best forecasts are a surprise, there is a trade-off between the information contained in a forecast and its chances of being correct. A forecast that relies on rather safe general principles will probably not provide any useful information. For example, 'in the future some new chemical products will be toxic' is rather a safe bet, but does not tell us anything useful. In contrast, 'future chemical products of formulation x will cause reproductive impairment of cyprinid fish', is more risky and, if correct, gives us highly useful information.
2. **Keep it general and vague**. General trends are easier to forecast than are specific events, and ambiguous forecasts are safest of all, as astrologers know well. It is also best not to make a concrete forecast of what will happen, but instead to describe several possible outcomes that may come about.
3. **Keep it limited**. If a forecast is hedged around with qualifications and limitations it will be safer.
4. **Keep it close to the present**. The near future is easier to forecast than events some distance into the future.
5. **Keep it simple**. The fewer the parameters in a forecast the more secure one can generally be.

More seriously, environmental toxicologists and chemists, and certainly the risk assessors who use information generated by these professionals, need to be clear about the type of phenomena that they are measuring. They need to use appropriate forecasting tools, and be aware of the limitations of both the tools and the limits to forecasting that affect the phenomena in which they are interested.

Disciplines differ in their forecasting accuracy and precision along a continuum. Some, like economics, sociology and ecology provide relatively weak forecasts because they deal with volatile and unstable phenomena that are subject to chance and chaos. This does not mean that these disciplines are incapable of forecasting anything, only that they cannot forecast everything we would like them to. Practitioners in other disciplines, such as biochemistry and toxicology, are fortunate in that forecasts can often be based on more stable parameters. If a choice were available then environmental toxicologists and chemists would be wise to make forecasts within stable scientific domains. If they must use phenomena in unstable domains, then they should be aware of the forecasting limitations and not expect to achieve the impossible.

Rescher (1998) classifies humans as predictability **believers**, predictability **sceptics** and predictability **cautionists**. Predictability believers, like the Stoics of antiquity, hold the view that the future can be forecast accurately if we can only uncover the natural laws that apply to the phenomena in which we are interested. Predictability sceptics believe that either the limitations of the human mind or the inherent complexities of nature prevent accurate forecasts. Predictability cautionists take a middle road. They acknowledge the difficulties and obstacles involved in accurate forecasting, but also recognize that some forecasting strategies have value. This appears to be the only pragmatic approach available to those who wish to forecast the fate, behaviour and effects of chemicals in nature. A belief in deterministic forecasting rules flies in the face of current views on chaos and complexity (although who is to say that these views may not change in the future!). On the other hand, if there is no rational basis for forecasting the future fate and effects of chemicals then we might just as well use voodoo or astrology.

How do we cope with our inability to forecast as accurately as we might wish? Rescher (1998) suggests three main strategies: inquiry, insurance and planning. In those areas of environmental toxicology and chemistry where our ignorance, rather than chance and chaos, prevents us from making correct forecasts we should endeavour to discover more. This will usually be a medium- to long-term strategy and will not yield immediate improvements in our forecasting skill. Insurance involves making a small, assured sacrifice in order to avoid a large potential loss. In the context of environmental toxicology and chemistry, monitoring programmes can provide some insurance against major toxic catastrophes. Both the inspection of industrial facilities and the monitoring of biology and chemistry in the natural world provide essential information on whether our forecasts of contaminant sources, fate and effects are sufficiently accurate. Finally, rational planning for the future does not require accurate forecasts (although clearly they would be very useful). Even if we naively assume that tomorrow will be the same as today, we can plan to adapt if this is not the case. Rapid feedback systems within environmental regulatory frameworks are essential if such planning is to remain flexible enough to deal with an unknown future.

REFERENCES

Ahlburg D and Land K (1992) Population forecasting: guest editors' introduction. *International Journal of Forecasting*, **8**, 289–298.

Caswell H (1996) Demography meets ecotoxicology: untangling the population level effects of toxic substances. In *Ecotoxicology: A Hierarchical Treatment*, Newman M and Jagoe CH (eds), Lewis, Boca Raton, FL.

Caulkin S (1995) Chaos Inc. *Across the Board*. July/August 1995, 33–36.

Chapman P, Crane M, Wiles JA, Noppert F and McIndoe EC (1996) *Asking the Right Questions: Ecotoxicology and Statistics*, SETAC-Europe, Brussels, Belgium.

Crane M (1997) Research needs for predictive multispecies tests in aquatic toxicology. *Hydrobiologia*, **346**, 149–155.

Crane M and Newman MC (1996) Scientific method in environmental toxicology. *Environmental Reviews*, **4**, 112–122.

de Fur P, Crane M, Ingersoll C and Tattersfield L (1999) *Endocrine Disruption in Invertebrates: Endocrinology, Testing and Assessment*, SETAC Press, Pensacola, FL.

Friedman M (1953) The methodology of positive economics. In *Essays in Positive Economics*, University of Chicago Press, Chicago IL, pp. 3–43.

Gleick J (1987) *Chaos: Making a New Science*. Penguin Books, New York.

Horgan J (1996) *The End of Science*. Little, Brown, London, UK.

Keyfitz N (1981) The limits of population forecasting. *Population and Development Review*, **7**, 579–593.

Lawton JH (1989) Food webs. In *Ecological Concepts*, Cherrett JM (ed.). Blackwell Scientific Publications, Oxford, UK, pp. 43–78.

Matthews RA, Landis WG and Matthews GB (1996) The community conditioning hypothesis and its application to environmental toxicology. *Environmental Toxicology and Chemistry*, **15**, 597–603.

May RM (1974) Biological populations with nonoverlapping generations: stable points, stable cycles and chaos. *Science*, **186**, 645–647.

May RM (1989) Levels of organization in ecology. In *Ecological Concepts*, Cherrett JM (ed.), Blackwell Scientific Publications, Oxford, UK, pp. 339–363.

McNees SK and Ries J (1983) The track record of macroeconomic forecasts. *New England Economic Review*, November/December, 5–18.

Newman MC (1995) *Quantitative Methods in Aquatic Ecotoxicology*. Lewis, Boca Raton, FL.

Pagels H (1988) *The Dreams of Reason*. Simon & Schuster, New York, NY.

Peters RH (1991) *A Critique for Ecology*. Cambridge University Press, Cambridge, UK.

Pindyck R and Rubinfeld DL (1976) *Econometric Models and Economic Forecasts*. McGraw-Hill, New York, NY.

Rescher N (1998) *Predicting the Future: An Introduction to the Theory of Forecasting*. State University of New York Press, Albany, NY.

Sherden WA (1998) *The Fortune Sellers*. Wiley, New York, NY.

Smith S and Sincich T (1992) Forecasting state and household populations. *International Journal of Forecasting*, **8**, 495–508.

Stebbing ARD, Travis K and Matthiessen P (1993) *Environmental Modelling— The Next Ten Years.* SETAC-Europe, Brussels, Belgium.

Stoto MA (1983) The accuracy of population projections. *Journal of the American Statistical Association* **78**, 13–20.

Woudenberg F (1991) An evaluation of Delphi. *Technological Forecasting and Social Change,* **40**, 131–146.

Zeeman MG (1995) Ecotoxicity testing and estimation methods developed under Section 5 of the Toxic Substances Control Act (TSCA). In *Fundamentals of Aquatic Toxicology,* 2nd edn, Rand GM (ed.), Taylor and Francis, pp. 703–715.

3

Uncertainty in Toxicological Predictions: the Bayesian Approach to Statistics

ANTHONY O'HAGAN

Department of Probability and Statistics, University of Sheffield, Sheffield, UK

3.1 INTRODUCTION

This is the only chapter in this book that is not written by a chemist, toxicologist or biologist, but by a statistician. The author's particular field is Bayesian statistics, its theory and applications. In recent years, the applications of Bayesian statistics by the author have increasingly tended to be in the area of environmental statistics.

The science of statistics plays an important role in a huge diversity of other disciplines, for example to help solve problems in telecommunications, the law, nuclear protection, the water industry, education, pollution monitoring, auditing, power generation, medicine, even social history! Statistical analysis is evident in many of the other chapters in this book. Like other sciences, statistics has been experiencing enormous growth, and the power and flexibility of modern statistical techniques could not have been dreamt of when the author was a student (yet the methods learnt then are predominantly those *still* taught in service courses to students of other disciplines). A major driving force behind this growth over the last few decades has been the development of the Bayesian approach to statistics.

The aims of this chapter are to explain, in quite simple terms, what Bayesian statistics is (including how it differs from the more traditional approach), and to indicate how Bayesian methods might be beneficial in forecasting the fate and effects of toxic chemicals.

3.2 MODELS

To motivate what follows, it is useful to begin by discussing models. Much of modern environmental science involves modelling of complex physical, chemical and biological processes. In environmental toxicology, we might consider the consequences of a particular quantity of some chemical entering a river, the atmosphere or a landfill site. The objective might be to predict the

Forecasting the Environmental Fate and Effects of Chemicals. Edited by Philip S. Rainbow, Steve P. Hopkin and Mark Crane.
© 2001 John Wiley & Sons Ltd

effects on organisms coming into contact with the chemical, or its dispersion through the medium and eventual fate. In order to do this, the toxicologist must use the best available scientific knowledge to link the initial entry of the chemical into the medium with the eventual consequences. Whatever this link looks like, we can think of it as a model.

3.2.1 MODELS IN GENERAL

It is possible to represent more or less any model mathematically by an equation

$$y = f(z, \theta)$$

The symbols in this equation have the following meanings:

1. y is the quantity to be predicted.
2. z represents known factors like the quantity of chemical, river flow or wind speed. It is not intended to be just a single number, and in practice will typically be a collection of numbers (mathematically, a vector).
3. θ represents unknown parameters which appear in the model, and will also typically be a collection of numbers.
4. f represents the mathematical relationship which the model establishes between y and (z, θ). It is a mathematical **function**. Given values for its arguments z and θ, the function yields a value which the model says equals y.

To further clarify the ideas, two examples follow.

3.2.2 ATMOSPHERIC DISPERSION MODELS

For a chemical released into the atmosphere, perhaps from a factory chimney, we may wish to predict what concentrations will be deposited at various distances from the source. So y might represent deposition at a particular location. Scientific knowledge and theories will be used to build a model for atmospheric dispersion, which can predict y if we know the relevant circumstances, such as: (1) stack height, (2) emission rate or volume, (3) wind speed and direction, (4) relevant geographical features (hills, etc.), (5) deposition rate for this chemical. The first four of these might well be known in a particular application, and will constitute the z part of the equation. On the other hand, the rate at which any chemical will drop out of the atmospheric plume will not usually be known exactly, and this will therefore be an unknown parameter. It will thus be (all or part of) θ.

The model itself, and the computations it performs given z and θ, is represented by the function f. It may be a very complex series of computations arising from highly sophisticated mechanistic modelling, programmed into a complex computer code. Or it could be a much simpler process which can be calculated by hand. In the case of a computer code, we would call z and θ the inputs to the program, and then $f(z, \theta)$ is the output, which in this case is the predicted deposition at a particular location.

3.2.3 TOXICITY MODEL

The second example involves more empirical modelling. We wish to predict some measure of toxicity for a chemical based on its structural properties. In a model for a quantitative structure–activity relationship (QSAR), the toxicity measure (y) is represented through a statistical relationship (f) in terms of one or more structural properties (z) and some unknown regression parameters (θ).

Using experimental data, the model is fitted statistically, meaning that the parameters θ are estimated. But θ is still technically unknown, because no amount of data will allow the parameters to be determined exactly.

3.2.4 UNCERTAINTY IN MODELS

By definition, θ is uncertain, but uncertainty in models is by no means confined to θ. It has been established above that z represents known quantities, but in the two examples the quantities that were identified as comprising z may not always be known. Consider the atmospheric dispersion example. Suppose that an unplanned release of a toxic chemical is known to have taken place from a plant. The stack height, wind direction and speed, and geography are known, but we may not know the magnitude of the release. Indeed, we may wish to use the model 'in reverse' to estimate this from observed depositions in the vicinity. This is called calibration of the model (see Kennedy and O'Hagan 1999).

In a different application of the atmospheric dispersion model, it may be required to assess the safety of a plant in terms of the probability that deposition at a given location exceeds a threshold. In this context, both the quantity/rate of release and the wind speed/direction can be regarded as random. The idea is to assess the probability that release levels and wind properties combine to produce an unacceptable level of deposition. This kind of problem, where we assess the uncertainty in a model's output as a result of uncertainty about inputs, is often called **uncertainty analysis** (see O'Hagan and Haylock 1997).

In the toxicity model, there may be a chemical for which only some of the structural properties have been measured. Thus for any model there may be contexts in which (at least some components of) x should be regarded as uncertain. An alternative way to express this is that the division of the model's inputs into known x and unknown θ is not rigid, but depends on context.

Moving on, the other symbol in our symbolic model is f. The postulated model/function f, no matter how sound the science on which it is based, is almost certainly wrong. We simply do not know the true functional relationship between y and (z, θ). So there is also uncertainty in f.

Finally, even if the model is perfectly correct and we know all the inputs, z and θ, it will still not predict y with perfect accuracy. The quantities we wish to estimate are inherently random. In the atmospheric dispersion model, for instance, the actual deposition will differ from even the best prediction due to intrinsic randomness. This should really be recognised explicitly by elaborating the symbolic model to

$$y = f(z, \theta) + e$$

where e is a random noise/error term. The random error term will generally be a part of an empirically based model like the toxicity example, but builders of 'science based' models all too often forget this basic point. More discussion of uncertainty in models may be found in Barnett and O'Hagan (1997).

3.2.5 THE UNCERTAINTY SPECIALISTS

So there is uncertainty everywhere! Many of these uncertainties are very difficult to assess or quantify. Nevertheless, it is important to do so. To pretend that an estimate is more accurate than it really is may be quite misleading. In the context of environmental protection, it can be downright dangerous.

This is really where the statistician comes in. It may be said that the job of a statistician is all about measuring uncertainty. Indeed uncertainty is the stock-in-trade of a statistician. Environmental toxicology and chemistry needs uncertainty specialists, because the field is afflicted with uncertainty to at least the extent that other disciplines are.

This book is about forecasting—forecasting the fate and effects of toxic chemicals. A forecast or a prediction is an estimate of something uncertain. It happens to be an estimate which is uncertain in the particular sense that it is in the future, but that makes it intrinsically no different from any other estimation problem. And because it is uncertain, a single value is too simplistic. No estimate is complete without an indication of its accuracy, a measure of its uncertainty. So bring in the uncertainty specialists.

3.3 BAYESIAN STATISTICS

3.3.1 AN ALTERNATIVE PHILOSOPHY

Readers who have attended a course in statistics were almost certainly taught the subject from the perspective of frequentist statistics (also sometimes misnamed classical statistics). Frequentist statistical procedures are built on the three inference tools of unbiased estimates, confidence intervals and significance tests. A statistical analysis in which the words 'unbiased', 'confidence' or 'significance' appear, or where a 'P-value' is reported, is a frequentist analysis. To this day, the vast preponderance of statistical analyses performed in practice are frequentist. All the industry standard statistical software packages (including SPSS, SAS, Minitab and BMDP) carry out frequentist methods.

Yet statistical analyses performed under a completely different paradigm, Bayesian statistics, are beginning to appear in a great variety of application areas. Historically, the Bayesian approach is actually older than the frequentist, but it is only since the 1960s that Bayesian statistics has developed its own body of theory and techniques to rival the frequentist school. It has been steadily gaining ground over the latter part of this century within the statistical profession. A sizeable and growing proportion of academic statisticians now advocate the Bayesian approach.

Bayesian statistics is built on a completely different philosophy than the frequentist approach. Paradoxically, in simple problems the two approaches often seem to coincide. A Bayesian statistician and a frequentist may do identical calculations, and report their inferences in apparently similar terms, yet the interpretations of these inferences are fundamentally different. In more complex problems, the two approaches can produce quite different answers.

3.3.2 PRIOR AND POSTERIOR DISTRIBUTIONS

In order to see how the approaches differ, and to appreciate what the Bayesian approach has to offer, it is necessary to understand at least the rudiments of Bayesian statistics. The name 'Bayesian' derives from the way Bayesian statistics uses **Bayes' theorem**. (The theorem is due to the Reverend Thomas Bayes, and was published in 1763, two years after his death.) Bayes' theorem, as used in Bayesian statistics, says

$$p(\theta|x) = k\,p(\theta)\,p(x|\theta).$$

where θ represents unknown parameters and x represents data. The essence of all statistical methods is the attempt to use data (x) to learn about the unknown parameters θ. The symbol p denotes a probability distribution (technically, it usually represents a probability density function). The vertical bar indicates a conditional probability and separates the variable whose distribution is being described (to the left of the bar) from a variable whose value is to be assumed known ('conditioned' on, to the right side of the bar). Here, k is a constant whose role will be explained shortly.

Thus, $p(\theta)$ is the probability distribution of the unknown parameters, whereas $p(\theta|x)$ is the probability distribution of θ when the value of x is **given**. The interpretation of these two distributions is important. $p(\theta|x)$ is the distribution that θ will have **after** we learn the value of x (i.e. **given** x) and is called the **posterior** distribution. $p(\theta)$ is the distribution that θ has **before** we observe x and is called the **prior** distribution. Bayes' theorem says how we should **learn** from the data, because it tells us how to convert the prior distribution to the posterior distribution.

The way that this learning occurs is basically that the prior distribution $p(\theta)$ is multiplied by the distribution $p(x|\theta)$ of the data given θ. The constant k is then chosen so that the product integrates to 1, which is a property which any probability distribution must have. This simple process is quite natural. For any given value of θ, $p(\theta)$ will be high if this value of θ is relatively probable on the basis of the information prior to observing x (the **prior information**), and if $p(\theta)$ is low then this value of θ is relatively unsupported by prior information. A similar interpretation applies to $p(x|\theta)$. If it is high then the data which have actually been observed would have been relatively likely if this is the true value of θ. If $p(x|\theta)$ is low, then the observed x would have been relatively improbable given this value of θ, and so suggests that this value of θ is relatively unsupported by the data.

Broadly speaking, if a value of θ is well supported by both information sources, the prior information and the data, then the product $p(\theta)p(x|\theta)$ will be large. But if θ is poorly supported by **one** of these information sources, so that either $p(\theta)$ or $p(x|\theta)$ is small, then $p(\theta)p(x|\theta)$ will be smaller. Clearly, a θ which is not well supported by **either** information source gets an even lower $p(\theta)p(x|\theta)$. Since the constant k is applied to $p(\theta)p(x|\theta)$ for all values of θ alike, the same considerations apply to $p(\theta|x)$. The posterior distribution will be high for a θ supported by both information sources, lower if θ is well supported by only one source and lower still if it is supported by neither.

Obviously the precise value and shape of $p(\theta|x)$ depends on the precise values and shapes of $p(\theta)$ and $p(x|\theta)$, but the above guidelines give a good idea of how Bayes' theorem accords with common sense. The following example may help to reinforce the reader's understanding.

3.3.3 EXAMPLE: CONCENTRATION RESPONSE

A standard way of assessing toxicity of a chemical is to expose some test organisms to different concentrations and to observe what proportion of those organisms are affected. Consider a part of such an experiment associated with one concentration. (The same principles apply when assessing the whole response curve, or summaries of it like EC_{50}, but they can be seen most clearly in this very simple case.)

To be more specific, imagine that the chemical has a certain concentration in water, and the test organism is a species of fish. The (single) unknown parameter θ is the probability that a single fish is affected by the chemical at this concentration. Based on prior information (which might include some observations made under less controlled conditions, for instance) the experimenter believes that the most likely values for θ are around 0.2, but that values from near zero to 0.5 or more cannot be ruled out. This prior information is reflected in the prior distribution $p(\theta)$ shown as a short-dashed line in Figure 3.1.

In the experiment, 25 fish are exposed and 10 are observed to be affected. The data clearly support values of θ around 0.4 and this is confirmed by $p(x|\theta)$, which is shown with long dashes in Figure 3.1. When these two curves are multiplied together, then rescaled by k so that the area under the product is 1, the resulting posterior distribution is shown as a solid line in Figure 3.1.

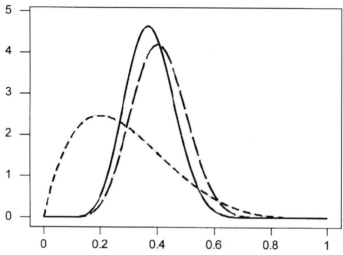

Figure 3.1.   Prior $p(\theta)$ (short dashes), likelihood $p(x|\theta)$ (long dashes)and posterior $p(\theta|x)$ (solid) for concentration response example.

The synthesis of the two information sources can be seen in the fact that the most probable value of θ is now 0.367, which is between the peaks at 0.2 and 0.4 suggested by the prior information and the data. But it is clearly much closer to 0.4 than 0.2. This is because the data represent a stronger information source. The strength of information is reflected in the width of the corresponding distribution curve (or in statistical terms, its variance). The prior distribution $p(\theta)$ is broad and flat, giving support to a wide range of θ values. As such, it is not very strong information. The $p(x|\theta)$ curve is narrower (and higher, so that the area under it is again scaled to 1), and so this is a stronger information source. Had our experiment used more fish, $p(x|\theta)$ would have been tighter still. The posterior distribution $p(\theta|x)$ is thus closer to $p(x|\theta)$ than to $p(\theta)$. It is also even narrower, as would be expected since it gains the strength of **both** information sources.

Although the prior information suggested that values of θ between 0 and 0.2 were quite likely, these are simply ruled out by the data. Values above 0.5 are not ruled out by either constituent, but are not strongly supported by either, and become still less supported by $p(\theta|x)$.

In this example, the prior information has only moderated the data information slightly. It is worth noting that the Bayesian learning process is dynamic. If tomorrow we observed the results from another tank of fish with the same concentration level, we could incorporate this new information by employing Bayes' theorem again. Today's posterior distribution would become tomorrow's prior, and we would have a new $p(x|\theta)$. This prior distribution would be much stronger, and would have accordingly more effect on the new posterior.

Figure 3.1 is called a triplot. Looking at triplots can be a useful way to gain appreciation of how Bayes' theorem works. There is a free software package called First Bayes, which is designed to help with learning elementary Bayesian statistics, and in particular includes the facility to view triplots. It is available from http://www.shef.ac.uk/~st1ao. However, it assumes a little more knowledge of (Bayesian) statistics than would be gained just from reading this chapter.

3.3.4 THE NATURE OF PRIOR INFORMATION

The principal difference between the Bayesian and frequentist approaches to statistics lies in the Bayesian's use of prior information. The distribution $p(x|\theta)$ is used in both approaches. When viewed as a function of θ for the fixed observed x (as in Bayes' theorem), it is called the **likelihood** and plays a major role in frequentist statistics. The frequentist maximum likelihood estimator, for instance, is just the value of θ maximizing $p(x|\theta)$ (e.g. 0.4 in Figure 3.1). But whereas all frequentist inference is based on $p(x|\theta)$, all Bayesian inference is based on the posterior distribution $p(\theta|x)$, which Bayes' theorem says is proportional to the product of the likelihood and the prior. The use of a prior distribution is the crucial extra step in a Bayesian analysis.

It is therefore worth focusing on the prior information a little, to understand what it comprises and how it is converted into the prior distribution $p(\theta)$.

As a crude generalization, the data x in any statistical analysis can be characterized as being well defined, clearly structured and objective. In contrast, prior information is generally based on less well defined or quantified experience, loosely structured and, to a greater or lesser extent, subjective.

Formally, the prior information comprises everything that you know apart from the data x. Before considering what 'everything' means here, we should discuss who 'you' might be. Because each person's prior knowledge will in principle differ from any other person's, each will have their own prior distribution $p(\theta)$, and hence their own posterior distribution $p(\theta|x)$. Herein lies the subjectivity of a Bayesian analysis, an issue that has provoked substantial and often heated debate. Strictly, 'you' should be the person for whom the statistical analysis is being done. (This is often, but not always, different from the person actually doing the analysis.) 'You' might be a scientist wishing to report the findings of some experiment, or a regulatory agency needing to decide on pollution controls.

'Your' prior information needs to be coded in the form of a probability distribution $p(\theta)$, describing what 'you' believe about θ based on that information. This is a difficult task in general. The process by which another person (e.g. the statistician) draws out and codes 'your' knowledge is called **elicitation** (although it can equally represent how 'you' analyse 'your' own prior knowledge). Elicitation has been studied by both psychologists and statisticians (see Kadane and Wolfson 1996 and O'Hagan 1998). A fascinating collection of essays ranging from the discursive to the technical will be found in Wright and Ayton (1994).

Obviously, we wish the prior distribution to reflect accurately 'your' genuine prior knowledge, not any of the unconscious prejudices, biases or superstitions of which we can all be victims. In fact, when the word 'subjectivity' causes concern, it is not usually a worry about different people genuinely having different information; that is inevitable, and genuine information should legitimately be exploited in any statistical analysis. The concerns are because 'subjectivity' also brings connotations of the irrational — prejudice, superstition, etc. Well-elicited prior information should be 'objective' in the sense of minimizing such influences and focusing on genuine knowledge.

Much more discussion of subjectivity and principles of measuring or eliciting probability distributions coherently may be found in O'Hagan (1988). This book is written to be as non-technical an introduction as possible to the probability ideas needed for Bayesian statistics, and that leads us into the next subsection.

3.3.5 THE NATURE OF PROBABILITY

When probability is taught at an elementary level, it is almost invariably defined in terms of limiting relative frequency. Consider the toss of a coin. According to this definition of probability, to say that the probability of heads is one-half, i.e. $Pr(heads) = 0.5$, means that if the coin is tossed over and over again the ratio of number of heads obtained to number of tosses (the 'relative frequency' of heads) would converge to 0.5 as the number of tosses goes to infinity.

Unfortunately, this definition just does not apply for all sorts of events which are simply not repeatable, and yet for which it seems natural to wish to use probabilities. Consider horse racing. One would imagine that betting on races is governed by probabilities and indeed it is, but they cannot be the conventional frequency probabilities. When a bookmaker offers odds of 10 to 1 against horse A winning in a certain race, he clearly thinks that there is a relatively small chance of A winning. If you decide to bet on A, then presumably you think the chance is rather better. But this race will never be repeated. Horse A has never run before against the same collection of other horses, and never will again. This already makes the situation unique, even if we do not ask for the race to be run also in the same conditions on the same course. The event of horse A winning this race is a one-off, to which frequency probability does not apply, yet most people would presumably accept that the 'chance' should be measured somehow by some kind of probability.

Here is another instance. Refer back to the concentration response example of subsection 3.3.3. The event that the proportion θ of fish affected by **this** chemical at **this** concentration is less than 0.2, say, is another one-off. It either is or is not less than 0.2, but the experimenter is uncertain and most people in that position would be willing to accept that this uncertainty should be measured by a probability. If the experimenter believed there was a 'good chance' that θ is less than 0.2 then this should somehow translate into a large probability.

There is such a definition of probability. It is called the personal definition. A personal probability for an event or proposition is simply a measure of a person's degree of belief in that event happening or that proposition turning out to be true. Under this definition, when a coin is to be tossed, the assertion that $Pr(\text{heads}) = 0.5$ means that, for the person making this judgement, the probabilities of heads and tails are equal. They have equal degrees of belief in the coin ending heads up or tails up, and so assign each a probability of one-half.

Bayesian statistics and the personal definition of probability clearly make a good combination. As in the concentration response example, the unknown parameters which statistical methods try to make inferences about are typically one-offs, and so it is not possible to express uncertainty about them using frequency probabilities. Prior knowledge is also a personal, subjective thing, and the personal definition of probability also fits with that context. Frequentist statistical theory is based on the frequency interpretation of probability, while Bayesian statistics employs personal probability.

The reader may wonder how any sensible scientific formulation of probability can be built on such a definition. O'Hagan (1988) discusses this in as non-technical a way as possible. Users of Bayesian methods are convinced that personal probability is a viable formulation of probability, and can be used effectively in practice. Those who reject Bayesian methods are unconvinced, for a variety of reasons. For some, any element of subjectivity in science is anathema. Matthews (1998) argues forcefully that, on the contrary, subjectivity is an inevitable feature of science in practice. Others reject Bayesian methods because, while they accept that prior information exists and can be useful, they feel that it cannot reliably be expressed so explicitly as in a prior distribution.

3.3.6 THE NATURE OF STATISTICAL INFERENCES

This distinction has a fundamental effect on the way that the two approaches make statistical inference statements. Remember that a frequentist approach **cannot** make probability statements about θ, because such statements have no meaning as frequency probabilities. So what does it mean when a statistician says that a 95% confidence interval for θ is from 0.17 to 0.33, say? It does not, **cannot**, mean that the probability that θ lies between 0.17 and 0.33 is 0.95. Yet this is what almost everyone will interpret it to mean.

What it actually means is that the statistician has applied a rule for constructing confidence intervals, and this rule has the property that if it were applied infinitely many times under identical conditions then 95% of the intervals so constructed would contain θ, no matter what value θ may have. The actual interval constructed on this occasion is 0.17–0.33. We do not know whether this is one of the 95% which contains θ or one of the 5% which do not. At this point, the reader could be forgiven for wondering why a distinction is being attempted. Surely, if 95% of intervals contain θ, and if we have no idea whether this is one of those 95%, the probability that it is one of the 95% is 0.95, and so the natural interpretation (that the probability of θ being between 0.17 and 0.33 is 0.95) is correct? Unfortunately, no.

It is a simple matter to devise **another** rule for constructing confidence intervals, so that from the particular data we have observed on this occasion the rule would yield the same interval of 0.17–0.33, and yet with the property that only 1% of its intervals will contain θ when it is applied infinitely many times. So if the previous reasoning were right the probability of θ being between 0.17 and 0.33 is not only 0.95 but also 0.01 (or any other value we like, since a rule for making confidence intervals can be found to give **any** desired value).

A confidence interval does not, **cannot**, have the interpretation which users of statistical methods nearly always understand it to have. Exactly the same argument applies to significance tests. If the hypothesis that $\theta > 0.5$ is rejected at the 1% significance level, often written as '$P < 0.01$', this does not, **cannot**, mean that the probability that $\theta > 0.5$ is less than 0.01. What it means is that the statistician has applied a rule for testing with the property that if θ really does exceed 0.5, and if this test were applied infinitely many times in identical conditions, then it would (incorrectly) reject the hypothesis on only 1% of those occasions. It does not, **cannot**, say anything about the probability that $\theta > 0.5$ on **this** occasion. Indeed, frequentist statistics do not recognize the existence or meaningfulness of such a probability.

The explanation of all this is simple. The only probabilities used in frequentist statistics are those making up the distribution $p(x|\theta)$. These are probabilities for x, the data. And they are frequency probabilities, so they refer to what happens in infinite repetition of (the experiment which produces) the data. They are not probabilities for θ, and frequentist inference statements do not make probability statements about θ. No matter how much the phrasing of those statements may invite interpretation as giving probabilities for θ, they do not and **cannot** be interpreted that way. There are plenty of real examples of how such interpretation has led to serious scientific errors; see Matthews (1998).

The contrast with Bayesian methods could not be more stark. Bayesian inferences are derived from the posterior distribution $p(\theta|x)$. Therefore, they **do** constitute probability statements about θ, and because they condition on the data x they refer to probabilities based on the actual observed data, not on infinite repetitions that we did **not** observe. If a Bayesian method gives the result that the range from 0.17 to 0.33 constitutes a 95% posterior probability interval for θ, the interpretation is exactly what you would expect — the probability that θ lies between 0.17 and 0.33 is 0.95, based on the actual observed data. Likewise, if we are interested in whether θ exceeds 0.5, a Bayesian analysis will compute the posterior probability that this hypothesis is true, based on the actual observed data.

There is of course a price to pay for having inference statements which genuinely say what they appear to say. You have to think about the available prior information, and formally express it as a (personal) probability distribution. This has to be done carefully and conscientiously, if you wish to avoid the kind of unconscious biases which would merit the epithet 'subjective'. When the prior information is substantive, and would make a worthwhile addition to the data, this is certainly a non-trivial task which makes the Bayesian approach harder or more time-consuming to apply. Yet this is also when the rewards are greatest. When prior information is negligible in comparison with the quality of information coming from the data, there are generally good shortcuts to make a Bayesian analysis more straightforward. The inferences **may** then turn out almost identical to the frequentist ones, but will now have the more natural and useful interpretation.

3.3.7 SUMMARY OF THE BAYESIAN APPROACH AND ITS ADVANTAGES

This section has been long and has introduced many new ideas. It will perhaps be helpful to summarize the main points and principal advantages here.

1. Bayesian methods make use of more information, because the prior information is utilized as well as the data. They will thereby generally produce stronger conclusions from the same data. Another aspect of this is that good answers can typically be obtained from smaller experiments than a frequentist analysis would require.
2. Prior information is formulated in a prior probability distribution for the unknown parameters. This necessitates employing personal probability. Prior information is subjective, and needs careful elicitation to avoid bias, prejudice, etc.
3. Bayesian methods, therefore, typically require more work, but through utilizing more information will generally yield stronger conclusions. Recently, very powerful computational techniques have been developed which facilitate the calculation of Bayesian inferences in extremely complex problems. Because they operate on a probability distribution for the unknown parameters, these methods cannot be applied in frequentist analyses. Many people are being drawn to use Bayesian methods, in a great variety of application areas, primarily because solutions cannot be obtained by frequentist methods. See the technical references by Gilks *et al.* (1996) and Tanner (1996)
4. Bayesian inferences are derived from the posterior distribution, and have a more direct, natural and useful interpretation than their frequentist counterparts. They are also more flexible because we are not limited to analogues of the three frequentist modes of inference (unbiased estimation, confidence intervals, significance tests). There is a whole branch of study under the theme of summarization, which is concerned with expressing the posterior information in the most useful way (see O'Hagan 1994, Chapter 2). Another topic, decision theory, concerns devising inferences (or decisions) to answer any specific question optimally. A very readable treatment of Bayesian decision theory is Lindley (1980).
5. Bayesian statistics is based on a more sound and philosophically satisfactory theory. This is not particularly relevant to readers of this book, but it is primarily what drove the early developers of the modern Bayesian approach in the 1950s and 1960s. Howson and Urbach (1993) is a spirited affirmation of the Bayesian approach by two philosophers of science.
6. Finally, a personal perspective: for a consultant in a wide range of fields, the Bayesian approach helps to build a rapport with the client. The Bayesian requirement to elicit prior information means that it is necessary for a Bayesian

statistician to try to better understand the context of the data, and to take more interest in the client's knowledge than a frequentist statistician might. People appreciate that!

3.3.8 FURTHER READING

This section has tried to provide the most basic of introductions to how Bayesian statistics work, with emphasis on how it differs from the frequentist approach which will be familiar to the great majority of readers. Further reading which will teach Bayesian statistics properly and thoroughly, and yet be suitable for the general reader has not been readily available until recently. Berry (1996), however, has written a very readable book at a very elementary level. Naturally, it does not deal with some finer points and does not go into any analyses more complex than the very simple ones typically tackled in elementary statistics courses.

The best recommendation for the reader with a good background in mathematics is Lee (1997). Lee gives more technical insight and extends to more complex problems, while still being an elementary text. For the professional statistician, the two 'bibles' are Bernardo and Smith (1994) and O'Hagan (1994).

3.4 TWO SCENARIOS

This section of the chapter presents two reasonably realistic examples of how a Bayesian approach might help with the kinds of problems faced by environmental toxicologists and chemists. The first illustrates the use of a formal Bayesian analysis to augment some data with prior information in the form of expert knowledge. The second shows the complexity of environmental models, and the presence of various uncertainties which are almost impossible to quantify with frequency probability.

3.4.1 SAFETY FACTORS

Suppose that you wish to estimate the EC_{50} (with effect defined in a suitable way) of chemical C for species S. You do not actually have any data relating to C-on-S toxicity in the field, but you have some laboratory data. If we let θ be the EC_{50} in laboratory conditions and ϕ be an EC_{50} in the field, the question is how to estimate ϕ. The standard technique, is to apply a 'safety factor' to the estimate of θ. This may mean dividing the θ estimate by 10, 100 or 1000. These safety factor values are acknowledged to be rather rough and ready, and ad hoc, but are thought to provide a margin of safety in protecting the environment.

The true ratio $\beta = \theta/\phi$ is of course highly unlikely to be any of these commonly used values, but is unknown. One difficulty is that it is not really well defined. The true EC_{50} in the field is not a unique number, but will depend on the particular conditions found in the 'field' in question. Safety factors are in effect trying to set more or less an upper bound on β, so as to play safe. Conditions in the field should rarely result in a ratio β greater than the assumed safety factor.

The word 'rarely' suggests that we should really be using probabilities. We do not know what range of β values might apply in practice because we have no field data, but safety factors are arrived at by specialists assessing the best scientific opinion, and we can use the same process. However, it is suggested here that it would be better to formulate that knowledge probabilistically by eliciting a probability distribution $p(\beta)$ for β.

Having done so, suppose that we wish to identify the EC_{50} value ϕ_0 which will be safe 99% of the time, by which it is meant that $\Pr(\phi < \phi_0) = 0.01$. One obvious way is to identify the 99% point β_0 of the distribution of β, so that $\Pr(\beta > \beta_0) = 0.01$, and call this the safety factor. It has the kind of meaning that the original safety factor was supposed to have, but arrived at in a more thorough way. We would then obtain a ϕ_0 value by dividing the estimate of θ from the laboratory data by β_0. But this ignores the uncertainty in θ. The estimate $\theta*$ from the data is not the true value, and the uncertainty remaining in θ means that $\theta*/\beta_0$ is not the ϕ_0 we need, in the sense that $\Pr(\phi < \theta*/\beta_0) \neq 0.01$. There is a general principle here that the safety factor you need to use to achieve a particular level of safety in the field depends on the accuracy of the laboratory EC_{50} estimate.

A proper analysis of this problem should treat both θ and β as unknown parameters. We should elicit prior information about θ as well as about β (although the experimental data are likely to be much more informative about θ). The resulting prior distribution should be combined with the data information via Bayes' theorem, yielding a posterior distribution for θ and β. The implied posterior distribution for $\phi = \theta/\beta$ should now be computed, and hence the suitably 'safe' EC_{50} in the field, ϕ_0, can be found.

In this analysis we do not explicitly analyse the information which the experts use when thinking about safety factors. We just treated it as prior information and elicited the experts' beliefs about β. There might actually be some well-defined data which the experts were implicitly including in their prior information, of greater or lesser relevance to β. There might be some data on toxicity of chemical C on one or two species similar to S, including some limited field-study data but at lower concentrations. We could consider taking these data out of the prior information and including them explicitly in the analysis. However, now we would need to look at many more unknown parameters, such as the ratios of EC_{50}s between related species. The analysis now becomes potentially very complex, and would demand expertise in Bayesian modelling and computation.

3.4.2 SETTING ENVIRONMENTAL STANDARDS

Figure 3.2 is based on a diagram in Barnett and O'Hagan (1997), showing a typical chain from cost to benefit in setting standards.

In order to set defensible and realistic standards on environmental pollution, Barnett and O'Hagan argue that legislators must balance the cost to polluters

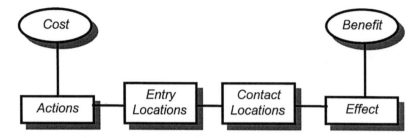

Figure 3.2. Links from cost to benefit.

(and thereby to society), of complying with the standard, against the benefit in terms of reduced ill effect on subject populations. The diagram shows that it is therefore necessary to understand:

1. How the actions of polluters affect the quantities of pollutants entering the relevant medium.
2. How this determines the concentrations of pollutants at entry locations.
3. The mechanisms for dispersal and ensuing concentrations at locations where individuals come into contact with pollutants.
4. Consequent effects on the subject populations.

It is important to recognize that uncertainty exists about all these links in the chain from cost to benefit. To understand these links, models are built as discussed in section 3.2. Models for links and sub-links are integrated into larger models. As discussed in section 3.2, there is uncertainty in all aspects of these models, and when we consider the problems of monitoring standards more uncertainties are introduced by the need for sampling.

The proper way to measure all these uncertainties is through probabilities. But, as we have seen, conventional frequency probability does not recognize many uncertainties as being describable by probabilities. For instance, suppose that in the link from entry location to contact location we could consider two alternative models for dispersal (e.g. atmospheric dispersal models). We should quantify the probability that model 1 is the more accurate (or 'correct') representation of the true dispersal behaviour, but this is not a repeatable situation. Such a probability can only be given meaning using personal probability, as acknowledged in the Bayesian approach.

Taking this example further, we can examine one last instance of where Bayesian methods can be of value. Suppose that we have quantified all the relevant uncertainties (and made use of statistical data, where available, to yield posterior distributions). We wish to know the costs and benefits associated with a proposed standard. These are, of course, uncertain, and we need to find their (posterior) probability distributions induced by the various uncertainties in all the models. In the example of safety factors there was a similar need to obtain the distribution of $\phi = \theta/\beta$ from those of θ and β. That is a simple computation, but the one we now consider will involve a very large number of uncertain parameters and model inputs, and an exact computation of the induced uncertainty on cost or benefit is out of the question.

This is a problem of 'uncertainty analysis'. A solution is to use simulation, often known as the Monte Carlo method. We randomly draw values in the computer for all the parameters and model inputs, drawing them from the probability distributions which we have derived for them. For each randomly drawn set of parameters, we run all the models and obtain outputs — random values of costs and/or benefits. The resulting samples of cost and benefit values represent (i.e. are effectively drawn from) the distributions of these quantities. To do this effectively we need a large sample, and that means carrying out a large number of runs of all the models. The computing time required is likely to be prohibitive, but even here Bayesian analysis can help. A Bayesian alternative to Monte Carlo uncertainty analysis exists and can dramatically reduce the number of model runs required (see O'Hagan and Haylock 1997).

It should finally be emphasized that it is not really practical to assess and quantify all the uncertainties present. That would be a huge and very difficult task. The nearest example known which comes even close to such analysis is the probabilistic risk assessment techniques used in the assessment of safety for nuclear waste disposal sites, see Bonano and Thompson (1993). **Many** man-years of work have gone into such assessments. Nevertheless, although the full analysis described above for setting environmental standards may be unattainable, the same principles apply to smaller, more manageable, versions of it, and to all kinds of similar analyses.

3.5 CONCLUDING REMARKS

This chapter has tried to clarify what the Bayesian approach to statistics is, and how it differs from the frequentist approach, and to give some tentative indicators of how it might beneficially be employed in forecasting the fate and effects of toxic chemicals. As explained at the outset, the author has no specialist knowledge of toxicology or any of the related fields. Probably the scenarios in section 3.4 are naive, but it is hoped that there is something in them that the reader will recognize as relevant to his or her own work.

Most of all, the author looks forward in the future to seeing (and perhaps playing a small part in) Bayesian methods realizing their power and potential in this field of application, as they have done in many others.

REFERENCES

Barnett V and O'Hagan A (1997) *Setting Environmental Standards: The Statistical Approach to Handling Uncertainty and Variation*. Chapman and Hall, London.

Bernardo JM and Smith AFM (1994) *Bayesian Theory*. Wiley, New York.

Berry DA (1996) *Statistics: A Bayesian Perspective*. Duxbury, London.

Bonano EJ and Thompson BGJ (eds) (1993) Special issue on probabilistic risk assessment of radioactive waste. *Reliability Engineering and System Safety*, **42**, 2–3.

Gilks WR, Richardson S and Speigelhalter DJ (1996) *Markov Chain Monte Carlo in Practice.* Chapman and Hall, London.

Howson C and Urbach P (1993) *Scientific Reasoning: The Bayesian Approach*, 2nd edn. Open Court, Chicago.

Kadane JB. and Wolfson LJ (1996) Experiences in elicitation. *The Statistician*, **47**, 1–20.

Kennedy MC and O'Hagan A (1999) *Bayesian Calibration of Computer Models.* Research Report 490/99, Department of Probability and Statistics, University of Sheffield.

Lee PM (1997) *Bayesian Statistics: An Introduction*, 2nd edn, Arnold, London.

Lindley DV (1980) *Making Decisions*, 2nd edn, Wiley, New York.

Matthews RAJ (1998) *Facts versus Factions: The Use and Abuse of Subjectivity in Scientific Research.* Working Paper 2/98, European Science and Environment Forum, Cambridge, UK.

O'Hagan A (1988) *Probability: Methods and Measurement.* Chapman and Hall, London.

O'Hagan A (1994) *Bayesian Inference*, volume 2B of *Kendall's Advanced Theory of Statistics.* Arnold, London.

O'Hagan A (1998) Eliciting expert beliefs in substantial practical applications. *The Statistician*, **47**, 21–35.

O'Hagan A and Haylock RG (1997) Bayesian uncertainty analysis and radiological protection. In *Statistics for the Environment 3, Pollution Assessment and Control*, Barnett V and Turkman KF (eds), Wiley, New York, pp. 109–128.

Tanner MA (1996) *Tools for Statistical Inference: Methods for the Exploration of Posterior Distributions and Likelihood Functions.* Springer-Verlag, New York.

Wright G and Ayton P (eds) (1994) *Subjective Probability.* Wiley, London.

4

Improving Inferential Strength of Exposure and Effect Forecasting: Working Outside the Box

MICHAEL C. NEWMAN* AND MORRIS H. ROBERTS, JR.

Virginia Institute of Marine Science, Gloucester Point, VA, USA

4.1 INTRODUCTION

Ecotoxicology and ecological risk assessment are new activities for which conceptual and technical frameworks are still evolving. Their present structures took form through the amalgamation of paradigms and techniques from classic mammalian toxicology, geochemistry, and ecology. Commonalities among these disciplines allowed this adoption and greatly accelerated progress. Because of dissimilar objectives and emphases among these fields, extra diligence was required during initial evaluation of borrowed paradigms and techniques. Unfortunately, practitioners understandably preoccupied by the immediacy of environmental stewardship have applied such diligence inconsistently. Adoption did not always include sufficient adaptation. Consequently the framework of ecotoxicological paradigms appears in some ways more chimeric (Kant 1934, Popper 1959) than harmonious. Key ecological risk assessment techniques produce accurate and precise data with which to answer the wrong questions, for example questions regarding effects on individuals but not populations (Barnthouse *et al.*, 1987, EPA 1991).

Both the science of ecotoxicology and the technology of risk assessment require periodic re-evaluation in order to cull away concepts and tools that no longer serve us well and to replace them with those that will. This is particularly critical in young fields that are evolving rapidly. Emphasis on re-evaluation is particularly important in ecotoxicology and ecological risk assessment because:

1. The consequences of inferior environmental stewardship are universally judged to be unacceptable.
2. The adoption of concepts and techniques from diverse fields produces inconsistencies that must be carefully resolved.

---

*Corresponding author

Forecasting the Environmental Fate and Effects of Chemicals. Edited by Philip S. Rainbow, Steve P. Hopkin and Mark Crane.
© 2001 John Wiley & Sons Ltd

3. US federal legislation wisely mandates periodic review of associated regulations to incorporate the most current science and technology.
4. The complex blend of scientific, technical and practical goals in these fields often results in muddled concepts and inappropriately applied technologies (Newman 1998).
5. Scientists tend to resist change (Barber 1961).
6. Technologists involved in environmental regulation are prone to become vested in a regulatory structure and standard methods that required much work to establish.

Our purpose here is to contribute to such re-evaluation by identifying five obvious areas where significant improvement can be made. Two involve exposure forecasting and the remaining three involve effect forecasting for toxic chemicals.

4.2 FIVE ILLUSTRATIONS

4.2.1 BIOAVAILABILITY ESTIMATION WITH THE AUC RATIO

Exposure is the simultaneous presence of a specific level of toxicant with a potential receptor (Ryan 1998). The amount of toxicant which is absorbed by the receptor and is available to have an effect, i.e. the effective dose, is predictable from exposure only if one has an accurate measure of bioavailability. If the exposure route is ingestion, bioavailability may be expressed as assimilation efficiency, the proportion of the total amount ingested that is incorporated into the organism. More specifically, bioavailability involves both the completeness of absorption of an ingested toxicant and the rate at which it is absorbed through the gut (Gibaldi 1991). Sometimes the added condition of 'and made available to interact with the site of action' is implied or stated relative to the concept of bioavailability (Newman 1998). Alternatively, digestion efficiency or absorption efficiency may be measured with slightly different methods and intents. Such estimates of bioavailability in ecotoxicology often entail a mass balance approach. Penry (1998) clarifies the difference between digestion efficiency (determined by difference in toxicant mass ingested and mass in faeces for a toxicant that is subject to digestion prior to uptake), absorption efficiency (determined by the difference in toxicant mass ingested and mass in faeces when no digestion is required before uptake, e.g. a tracer compound that is absorbed unmodified) and assimilation efficiency (determined by the difference in toxicant mass ingested and mass retained in the tissues of the organism).

In the context of bioavailability applied here, assimilation efficiency seems the most appropriate of these terms with the qualifier that the focus is presence in plasma or blood. For estimation of assimilation efficiency, a known amount of toxicant is ingested over time and assimilation efficiency estimated by the proportion of the administered dose retained after sufficient time has elapsed for gut clearance. Much effort is made to ensure that only inconsequential amounts of unassimilated toxicant remain in the gut, hepatopancreas or similar areas in the body. The presence of unassimilated toxicant biases estimates of assimilation efficiency upward. With this method, rapid elimination of any assimilated toxicant can bias estimates downward. Provided that these biases are minimal, a reasonable measure of oral bioavailability can be obtained for toxicants. It is often difficult to determine if these biases are minimal.

A common and efficient area under the curve (AUC) method from pharmacology can be, but has not been, generally applied in ecotoxicology to estimate bioavailability. Bioavailability is measured as the fraction of an administered dose reaching systemic circulation and available to interact at a site of action. A known dose is injected intravenously into an individual (100% bioavailable dose) and a curve of plasma concentration versus time after injection constructed by sampling blood over a time course. Then the same dose is given orally and a plasma concentration versus time curve constructed for this exposure route. The ratio of the areas under two curves is an estimate of bioavailability.

$$F = \frac{AUC_{oral}}{AUC_{iv}}$$

These calculations can be more involved if required. Ancillary constants such as the first-order rate constant for gut absorption and mean absorption time can be estimated with well-established methods (see Gibaldi 1991 and Newman 1995, 1998).

Despite its ease of application and advantages of avoiding the biases described above, the AUC method is rarely used in ecotoxicology. The AUC method was applied to estimation of antibiotic bioavailability to salmon (Horsberg et al. 1996, Martinsen et al. 1993a,b). It was applied for orally ingested chlorpyrifos by Barron et al. (1991) and methyl mercury by McCloskey et al. (1998).

McCloskey et al. (1998) measured methyl mercury bioavailabilities ranging from 14 to 55% in catfish (Ictalurus punctatus) receiving different amounts of food and demonstrated a clear relationship between ingestion rate and assimilation efficiency. More applications of this straightforward and useful technique would provide valuable information in risk assessments attempting to translate exposure to bioaccumulation or effect.

4.2.2 BIOACCUMULATION INVOLVING OSCILLATIONS

Ecotoxicologists modelling bioaccumulation assume that the equations below predict a gradual, monotonic increase in concentration until some steady-state concentration is reached. The differential equation for simple first-order bioaccumulation is the following,

$$\frac{dC}{dt} = k_u C_s - k_e C$$

Integration of this equation yields the simple first-order bioaccumulation,

$$C_t = C_s \left[\frac{k_u}{k_e} \right] [1 - e^{-k_e t}]$$

where C_s, C and C_t are toxicant concentrations in the source, organism, and organism at a specific time (t), respectively, k_u the uptake clearance rate, and k_e the elimination rate constant.

The model defined by these equations carries several assumptions. There is a toxicant source of constant concentration. There is instantaneous mixing in a single compartment with the probability of a molecule being eliminated from that compartment being independent of how long it has been in the compartment. The k_e and k_u are assumed to remain constant through time. However, there are likely as many situations as not in which one or more of these assumptions is violated. Lags occur between initial absorption and final elimination. The values of k_e and k_u vary with tidal, diurnal and annual cycles. Similarly, sources can be pulsed. The consequence of non-instantaneous mixing and cyclic changes in kinetics are time lags that can be included in this simple model to make it more realistic. The differential equation below includes a time lag (T) and the difference equation has a time step (τ) that can accommodate lags.

$$\frac{dC_t}{dt} = k_u C_s - k_e C_{t-T}$$

$$C_{t+\tau} = C_t + k_u C_s - k_e C_t$$

These minor changes to enhance realism produce major changes in bioaccumulation dynamics (Newman and Jagoe 1996). Depending on the magnitude of elimination (k_e) and lags/delays (T or τ) (Table 4.1), the predicted concentrations may either approach a steady-state concentration monotonically,

Table 4.1 Stability regions for the differential equation with time lag and difference equation for simple bioaccumulation

	Stability regions for equation	
Dynamics	Differential equation with lag	Difference equation
Monotonic increase to steady-state concentration	$0 < k_e T < e^{-1}$	$\tau < k_e^{-1}$
Damped oscillations to steady-state concentration	$e^{-1} < k_e T < \pi/2$	$\tau/2 < k_e^{-1} < \tau$
Diverging oscillations	$k_e T > \pi/2$	$k_e^{-1} < \tau/2$

oscillate slowly to a steady-state concentration or oscillate unstably through time. Only the first possibility is considered in our current use of bioaccumulation models for forecasting. This may be inappropriate as 'Omission of such time lags in this foundation model has drawn attention away from a potentially important model quality, equilibrium stability' (Newman and Jagoe 1996). Whether other delays or lags reduce oscillations (i.e. amplitude 'death by delay' of coupled limit-cycle oscillators (Strogatz 1998)) remains an unanswered question.

Why is this important? Many applications of bioaccumulation models assume a steady-state concentration. For example, the biocentration factor (BCF) concept and associated techniques assume that there is a steady-state concentration. If one considers equilibrium stability criteria (Table 4.1), it is clear that this may not always be a sound assumption. There may be a family of BCF values expected for bioaccumulation through time, not a single BCF value. Forecasting effect based on accumulation to a single critical body residue also may not be adequate if concentrations oscillate within the individual. A predicted limiting permissible concentration for species that are eaten by humans may also be suspect if some assumptions are violated.

4.2.3 QUANTITATIVE ION CHARACTER–ACTIVITY RELATIONSHIPS (QICARS)

Quantitative structure–activity relationships (QSARs) enjoy wide use for predicting bioaccumulation and effect of organic compounds. These empirical relationships rely on measures of molecular qualities such as lipophilicity, ionization or molecular topology to quantitatively predict effect or bioaccumulation. Such relationships are rare for predicting effects of metals. This is unfortunate as ecological risk assessment could be enhanced by reliable models for predicting effects on untested metals from known effects of tested metals. Such models would be useful in preliminary screening and in situations analogous to those in which QSARs are currently used.

Quantitative ion character–activity relationships (QICARs) can be generated under the supposition that effects of metal ions are correlated with aspects of metal-ligand binding chemistry. Like the lipophilicity of a non-polar organic compound, measures of metal-ligand binding tendencies can be used to predict metal effect. If accommodation of metal speciation is necessary, models can be developed under the general assumption that the metal ion is the active form of any metal (ion theory of Mathews 1904) and the specific assumption that the free ion is the most active of the ionic species (free ion activity model or FIAM (Campbell and Tessier 1996)). Fisher (1986) applied such an approach to predicting metal bioaccumulation in marine phytoplankton from the solubility products of the corresponding metal hydroxides, a measure of metal binding to intermediate ligands (i.e. ligands with O donor atoms such as hydroxyl and carboxyl functional groups). Newman and McCloskey (Newman and McCloskey 1996a, McCloskey *et al.* 1996) successfully generated QICARs for metal effects on bacterial bioluminescence. This work was extended to include QICARs for lethal effects to a soil nematode, *Caenorhabditis elegans* (Tatara *et al.* 1997, 1998). The general utility of the approach was demonstrated with a statistical analysis of QICARs for a wide range of effects and species (Newman *et al.* 1998).

4.2.4 TOXICANT EFFECT AS A FUNCTION OF INTENSITY AND DURATION OF EXPOSURE

Current measures of toxicant effect involve a design in which effects are measured after a specific exposure duration at a series of constant concentration or dose treatments. Some metric of effect such as a 96 h LC_{50} is generated from the resulting data and used as the standard measure for lethal effect in ecological risk assessments. If the concentration killing 50% of exposed individuals is judged to be too high for risk assessment, the slope of the concentration-effect curve is used to predict an effect level lower than 50% or the LC_{50} is divided by some arbitrary number such as 10.

Because both intensity (dose or concentration) and duration of exposure are important in predicting effect, the dose (concentration)-effect approach provides temporally restricted information for forecasting in ecological risk assessment. It provides predictions for only one duration. To gain more information about the influence of exposure duration, a series of tests may be carried out at different exposure durations or effects may be noted daily or more often in one test. However, as will be detailed below, there are several difficulties with these modifications of the dose (concentration)-effect approach.

Another approach, time-to-death or survival time analysis, is widely applied in other fields to quantify death through time (see Miller 1981, Cox and Oakes 1984 and Marubini and Valsecchi 1995 for a comprehensive explanation of these methods), but it is rarely applied in ecotoxicology (Newman 1995, 1998). This approach consists of periodically noting the time-to-death for individuals in various treatments (e.g. toxicant concentrations). Models are developed for survival time as influenced by covariates such as toxicant concentration, and hypothesis tests applied to determine which covariates have statistically significant effects on survival. Applicable methods range from non-parametric to semi-parametric to fully parametric. They produce more information than the dose (concentration)-effect approach from the same experimental design (e.g. 10 times-to-death observations for a tank of 10 fish instead of only one proportion at 96 h) with a significant increase in statistical power (Gaddam 1953, Finney 1964, Sprague 1969, Newman 1995).

Because high statistical power and the ability to predict effect at different exposure durations are important in ecological risk assessment, it is difficult to understand why survival time methods have not been used extensively in ecotoxicology. The reason seems to be linked to the past focus on controlling toxicant concentration discharged from point sources. However, the current application of effects data in forecasting consequences during risk assessments demands more from lethality data than can be provided by the conventional dose (concentration)-effect approach. Survival time methods provide readily applicable means to meet these emerging demands. These methods also generate survival information amenable to predicting population consequences of exposure, i.e. life-table analyses (Newman and McCloskey, in press). Application of survival time methods to ecotoxicological problems is illustrated in Dixon and Newman (1991), Newman (1995), Newman and Aplin (1992), Newman and Dixon (1996), Newman and McCloskey (1996b).

4.2.5 INDIVIDUAL EFFECTIVE DOSE OR TOLERANCE CONCEPT

Data generated with the dose (concentration)-effect methods just described are most often analysed by probit analysis. Probit analysis carries the assumption of a log-normal model for effect (proportion dying) versus dose (or concentration). The individual effective dose (IED), individual lethal dose or individual tolerance concept is the accepted explanation for this log-normal model (Finney 1947). There is a certain dose below which an individual will survive, but above which it will die. Described first in the context of dose (Bliss 1935), this concept can also be used relative to an individual effective concentration. This individual effective dose or concentration is an inherent characteristic of an individual although Finney (1947) did suggest that it could change for an individual with time. The distribution of these tolerances among individuals in a population is most often a log-normal distribution. By repetition instead of careful testing, this concept has evolved into a truism. However, the concept failed as an explanation in the only case to the authors' knowledge in which it was challenged (Berkson 1951). Further, the widespread application of the probit method to data for clones or uniform cultures (e.g. the *Daphnia* toxicity test or Microtox® bacteria) would be difficult to explain with the IED concept. What would be the source of variation in IED values among individuals?

An alternative explanation would treat death as a random process that occurs identically in all exposed individuals. Which particular individual dies at a dose or concentration is simply a matter of chance. The probability function for this process is best described by a log-normal model.

Two questions emerge as important at this point in the argument. First, which of the two explanations is correct? Second, is it important which is correct relative to forecasting effects of toxicants? Surprisingly, it remains untested which is the best explanation, although Newman and McCloskey (2000) suggest that the random process explanation seemed most likely for sodium chloride and pentachlorophenol lethality to mosquitofish (*Gambusia holbrooki*) and the individual effective dose was the most likely for stupefaction of zebra fish (*Brachydanio rerio*) with benzocaine. The importance of resolving this ambiguity can be demonstrated with a thought problem (Newman 1998). Assume that a population of individuals below an industrial outfall experiences a series of exposures to an 48 h LC_{50} for exactly 48 h each time. For simplicity, let covariates such as animal sex, size and age be identical for all individuals. The

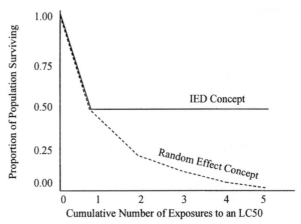

Figure 4.1 Consequences of repeated exposures for 48 h to an LC_{50} under the contrasting assumptions of the IED (solid line) or random effect (dashed line) concepts.

individuals are allowed to recover between exposures. As shown in Figure 4.1, the population would drop to 50% of its original size at the first exposure but drop only minimally thereafter if the IED concept were correct. The most sensitive individuals would be eliminated at the first exposure and the survivors would be sufficiently tolerant to live through subsequent exposures. But, if the random process concept were correct, the population would drop 50% each time it was exposed to the toxicant. Forecasts of realized effect after repeated toxicant exposure are very different based on these two unresolved explanations for the probit model. Current applications of results from probit analysis in ecological risk assessment activities arbitrarily adopt one explanation or the other. However, as demonstrated, predictions can be quite divergent.

4.3 CONCLUSIONS

The five examples detailed above support the premise that current paradigms and techniques used in ecotoxicology and risk assessment require careful re-examination. Such a re-evaluation would enhance our abilities to forecast exposure, dose and effects of chemical contaminants. To draw support for this argument from the five examples, the following improvements would result from such a reassessment:

1.   Accurate bioavailability data could be generated quickly for predicting realized dose in risk assessments by applying the well-established AUC method from pharmacokinetics.
2.   There may be instances in which it is invalid to apply a single BCF, critical body residue or limiting permissible concentration. Which instances can be identified with stability criteria (Table 4.1).

3. Prediction of metal effects could be made in situations in which specific information is missing, i.e. situations such as those in which QSARs are currently applied for organic compounds.
4. Toxicant effect can be more accurately predicted based on both exposure duration and intensity.
5. Consequences of toxicant exposure could be more reliably forecast if the underlying mechanism for the log-normal model (probit) were known. This was illustrated with a thought problem involving a repeated exposure scenario.

Clearly, there is practical value in carefully reviewing current paradigms and techniques. It would optimize the required review of federal regulations for periodic inclusion of the most current technology and best science. Some additional examples where our ability to predict consequences of exposure could be substantially improved are the following:

1. A reassessment of methods for predicting effects on populations (Barnthouse *et al.* 1987, Caswell 1996).
2. Development of methods to predict mixture effects.
3. Replacement of the flawed NOEC approach with regression-based methods (Stephan and Rogers 1985, Newman 1995).
4. Formalization of the expert opinion/weight-of-evidence approach by inclusion of Bayesian methods (e.g. Lane *et al.* 1987).
5. Replacement of current methods of statistical analysis of data sets with ' < DL' observations (Newman 1995).
6. Resolution of the controversy regarding the community redundancy theory versus rivet popper theory (Pratt and Cairns 1996).

Equally important, there is intrinsic value in reassessing techniques and concepts of any science. Failure to appreciate the value of periodic review and the falsification process stymies progress in any science. Review and formal testing of our current paradigms would accelerate the rate at which ecotoxicology matures as a science.

REFERENCES

Barber B (1961) Resistance by scientists to scientific discovery. *Science*, **134**, 596–602.
Barnthouse LW, Suter II GW, Rosen AE and Beauchamp JJ (1987) Estimating responses of fish populations to toxic contaminants. *Environmental, Toxicology and Chemistry*, **6**, 811–824.
Barron MG, Plakas SM and Wilga PC (1991) Chlorphyrifos pharmacokinetics and metabolism following intravenous and dietary administration in channel catfish. *Toxicology and Applied Pharmacology*, **108**, 474–482.
Berkson J (1951) Why I prefer logits to probits. *Biometrics*, **7**, 327–339.
Bliss CI (1935) The calculation of the dosage-mortality curve. *Annals of Applied Biology*, **22**, 134–307.

Campbell PGC and Tessier A (1996) Ecotoxicology of metals in the aquatic environment: Geochemical aspects. In *Ecotoxicology: A Hierarchical Treatment*, Newman MC and Jagoe CH (eds), CRC/Lewis, Boca Raton, FL, pp. 11–58.

Caswell H (1996) Demography meets ecotoxicology: Untangling the population level effects of toxic substances. In *Ecotoxicology A Hierarchical Treatment*, Newman MC and Jagoe CH (eds), CRC/Lewis, Boca Raton, FL, pp. 255–292.

Cox DR and Oakes D (1984) *Analysis of Survival Data*, Chapman & Hall, London.

Dixon PM and Newman MC (1991) Analyzing toxicity data using statistical models of time-to-death: An introduction. In *Metal Ecotoxicology: Concepts and Applications*, Newman MC and McIntosh AW (eds), Lewis Publishers, Chelsea MI, pp. 207–242.

EPA (1991) *Summary Report on Issues in Ecological Risk Assessment*, EPA/625/3-91/018 February 1991. NTIS, Springfield, VA.

Finney JH (1947) *Probit Analysis. A Statistical Treatment of the Sigmoidal Response Curve*, Cambridge University Press, Cambridge.

Finney JH (1964) *Statistical Method in Biological Assay*, Hafner, New York.

Fisher NS (1986) On the reactivity of metals for marine phytoplankton. *Limnology and Oceanography*, **31**, 443–449.

Gaddam JH (1953) Bioassays and mathematics. *Pharmacology Review*, **5**, 87–134.

Gibaldi M (1991) *Biopharmaceutics and Clinical Pharmacokinetics*, 4th edn, Lea & Febiger, Philadelphia, PA.

Horsberg TE, Hoff KA and Nordmo R (1996) Pharmacokinetics of florfenicol and its metabolite florfenicol amine in Atlantic salmon. *Journal of Aquatic Animal Health*, **8**, 292–301.

Kant I (1934) *Critique of Pure Reason*, Everyman Library, London.

Lane DA, Kramer MS, Hutchinson TA, Jones JK and Naranjo C (1987) The causality assessment of adverse drug reactions using a Bayesian approach. *Pharmaceutical Medicine*, **2**, 265–283.

Martinsen B, Horsberg TE, Varma KJ and Sams R (1993) Single dose pharmacokinetic study of florfenicol in Atlantic salmon (*Salmo solar*) in sea water at 11 °C. *Aquaculture*, **112**, 1–11.

Martinsen B, Sohlberg S, Horsberg TE and Burke M (1993) Single dose kinetic study of sarafloxacin after intravascular and oral administration to cannulated Atlantic salmon (*Salmo solar*) held in sea water at 12 °C. *Aquaculture*, **118**, 49–52.

Marubini E and Valsecchi MG (1995) *Analysing Survival Data from Clinical Trials and Observational Studies*, Wiley, New York.

Mathews AP (1904) The relation between solution tension, atomic volume, and the physiological action of the elements. *American Journal of Physiology*, **10**, 290–323.

McCloskey JT, Newman MC and Clark SB (1996) Predicting the relative toxicity of metal ions using ion characteristics: Microtox® bioluminescence assay. *Environmental Toxicology and Chemistry*, **15**, 1730–1737.

McCloskey JT, Schultz IR and Newman MC (1998) Oral bioavailability of methyl mercury to channel catfish, *Ictalarus punctatus*, *Environmental Toxicology and Chemistry*, **17**, 1524–1529.

Miller Jr RG (1981) *Survival Analysis*, Wiley, New York.

Newman MC (1995) *Quantitative Methods in Aquatic Ecotoxicology*, CRC/Lewis, Chelsea, MI.

Newman MC (1998) *Fundamentals of Ecotoxicology*, CRC/Ann Arbor Press, Boca Raton, FL.

Newman MC and Aplin M (1992) Enhancing toxicity data interpretation and prediction of ecological risk with survival time modeling: An illustration using sodium chloride toxicity to mosquitofish (*Gambusia holbrooki*). *Aquatic Toxicology*, **23**, 85–96.

Newman MC and Dixon PM (1996) Ecologically meaningful estimates of lethal effect on individuals. In *Ecotoxicology: A Hierarchical Treatment*, Newman MC and Jagoe CH (eds), CRC/Lewis, Boca Raton, FL, pp. 225–253.

Newman MC and Jagoe RM (1996) Bioaccumulation models with time lags: dynamics and stability criteria. *Ecological Modelling*, **84**, 281–286.

Newman MC and McCloskey JT (1996a) Predicting relative toxicity and interactions of divalent metal divalentmetal ions: Microtox® bioluminescence assay. *Environmental Toxicology and Chemistry*, **15**, 275–281.

Newman MC and McCloskey JT (1996b) Time-to-event analysis of ecotoxicity data. *Ecotoxicology*, **5**, 187–196.

Newman MC and McCloskey JT (2000). The individual tolerance concept is not the sole explanation for the probit dose–effect model. *Environmental Toxicology and Chemistry*, **19**, 520–526.

Newman MC and McCloskey JT (in press). Applying time-to-event methods to assess pollutant effects on populations. *Improving Risk Assessment with Time-to-event Models*, SETAC Press, Pensacola, FL.

Newman MC, McCloskey JT and Tatara CP (1998) Using metal-ligand binding characteristics to predict metal toxicity: quantitative ion character–activity relationships (QICARs). *Environmental Health Perspectives*, **106** (Suppl. 6), 1263–1270.

Penry DL (1998) Application of efficiency measurements in bioaccumulation studies: Definitions, clarifications, and a critique of methods. *Environmental Toxicology and Chemistry*, **17**, 1633–1639.

Popper KR (1959) *The Logic of Scientific Discovery*, Routledge, London.

Pratt JR and Cairns Jr J (1996) Ecotoxicology and the redundancy problem: Understanding effects on community structure and function. In *Ecotoxicology: A Hierarchical Treatment*, Newman MC and Jagoe CH (eds), CRC/Lewis, Boca Raton, FL, pp. 347–370.

Ryan PB (1998) Historical perspective on the role of exposure assessment in human risk assessment. In *Risk Assessment: Logic and Measurement*, Newman MC and Strojan CL (eds), CRC/Ann Arbor Press, Boca Raton, FL, pp. 23–43.

Sprague J (1969) Measurement of pollutant toxicity to fish. I. Bioassay methods for acute toxicity. *Water Research*, **3**, 793–821.

Stephan CE and Rogers JW (1985) Advantages of using regression to calculate results of chronic toxicity tests. In *Aquatic Toxicology and Hazard Assessment: Eighth Symposium, ASTM STP 891*, Bahner C and Hansen DJ (eds), American Society for Testing of Materials, Philadelphia, PA, pp. 328–338.

Strogataz SH (1998) Death by delay. *Nature (London)*, **394**, 316–317.

Tatara CP, Newman MC, Williams P and McCloskey JT (1997) Predicting relative toxicity with divalent ion characteristics: *Caenorhabditis elegans. Aquatic Toxicology*, **39**, 279–290.

Tatara CP, Newman MC, McCloskey JT and Williams PL (1998) Use of ion characteristics to predict relative toxicity of mono-, di-, and trivalent metal ions: *Caenorhabditis elegans* LC50. *Aquatic Toxicology*, **42**, 255–269.

5

Linking Pollutant Transport, Environmental Forecasting and Risk Assessment: Case Studies from the Geosphere

SIMON J. T. POLLARD AND RAQUEL DUARTE-DAVIDSON

Environment Agency National Centre for Risk Analysis and Options Appraisal, London, UK

5.1 INTRODUCTION

The study of the geosphere, and in particular of the soil matrix, has played a key role in improving our understanding of the behaviour of pollutants in the total environment. For example, advances in the understanding of metal and organic partitioning in soils have made important contributions towards advancing the analytical sciences and chemical risk-assessment methodologies. They have also helped to advance our understanding of pollutant treatment technologies and the environmental regulations that seek to control direct and indirect releases to the geosphere.

Environmental risk assessment has emerged over recent years as a widely accepted tool for environmental management in business, regulation and research (Calow 1998, European Environment Agency 1998). In many instances, the value of this activity has been in formalising existing judgements within an overall decision-making framework for the purpose of effectively managing risk. In this chapter, we present and examine the role of environmental fate and transport processes within the wider context of environmental risk assessment. Pollutant transport processes are viewed afresh as components of exposure assessment, set between the hazard identification and risk estimation stages within a classical risk-assessment structure. Case studies from the geosphere are used to illustrate the role of pollutant transport processes in informing regulatory decisions about the management of environmental risk.

5.2 POLLUTANT TRANSPORT PROCESSES

The law of the conservation of mass allows us to follow the fate and transport of pollutants by virtue of a simple mass balance. Within a system boundary (or

Forecasting the Environmental Fate and Effects of Chemicals. Edited by Philip S. Rainbow, Steve P. Hopkin and Mark Crane.
© 2001 John Wiley & Sons Ltd

between boundaries), whether at the global or field scale, the flow of pollutants can be expressed simply as

$$\text{Input rate} = \text{output rate} + (\text{decay rate}) + (\text{accumulation rate}) \tag{1}$$

The assumption of steady state (equilibrium) allows the accumulation rate to be set to zero. The conservative assumption of setting the decay rate (biotransformation or radioactive decay) to zero allows the input rate to be equated to the output rate. Finally, the assumption of homogeneity (i.e. that the distribution of pollutants throughout the system volume is uniform) completes the simplest view of the environment; that of the homogenous, 'complete mix box' or 'continuously stirred tank' (Masters 1991, Alloway and Ayers 1993).

This has been the starting-point for many approaches to the modelling of pollutant transport processes, (Schnoor 1996) and, as needs have dictated, progressive layers of complexity have been incorporated into environmental models. Examples include the inclusion of **storage** terms (lake phosphorus), **decay** terms (groundwater modelling of mineralizable organics, radionuclide decay) and the additional complexities with respect to mixing, including **advection** (riverine and free product flow), **diffusion** (bottom sediments) and **dispersion** (air quality modelling). These processes have been widely used to assess how, once released, pollutants are disseminated in the wider environment and, in particular, how they reach a point of **exposure**, be it a public supply borehole, point of inhalation downwind of a stack, or the site of inadvertent ingestion for a soil particle. Several texts have described pollutant fate and transport processes and modelling approaches in detail (Mackay 1991, Schnoor 1996).

5.3 ENVIRONMENTAL RISK ASSESSMENT

5.3.1 WHAT IS ENVIRONMENTAL RISK?

'Risk' is a term used to denote the probability of suffering harm from a hazard and embodies both likelihood and consequence. The hazard with which we are concerned, refers to the potential adverse effect posed by a source — a toxic substance or hazardous situation — and the effect represents the potential to cause harm. The actual harm that results from a risk relates to the **observable damage** that occurs once the hazard is realized and which is also referred to as the detriment, impact or response. Hazard, risk and harm are discrete terms and their interchangeable use is to be avoided. Environmental risk assessment is concerned with risks to and from the environment. Pollutant transport has most to do with the likelihood of environmental exposures from a chemical hazard.

As an illustrative example, consider one of the risks from methane gas emanating from an active landfill site. Methane gas represents an explosion (effect) hazard at certain concentrations in confined spaces. Methane poses a high risk where the likelihood of concentrations building up to the lower explosive limit are high (high probability, due to sufficient gas pressure, permeable strata, building ingress, etc.) and where there are occupied buildings in the vicinity (high consequences). The damage that can result from explosion risks includes loss of life, property and structural damage (harm).

5.3.2 ENVIRONMENTAL RISK ASSESSMENT

Risk assessment is a process for combining what is known, and what can be reasonably inferred about an exposure situation for the purpose of managing risk (Department of the Environment 1995, Calow 1998, Douben 1998, Hester and Harrison 1998). It typically involves answering the following questions which have become grouped into the classical stages of risk assessment (Box 5.1; Figure 5.1).

Box 5.1   Essential questions in environmental risk assessment

Question	Stage and definition
1. What hazards are present and what are their properties?	**Hazard identification**: Identification of the sources of the hazard and assessment of the consequences of the hazard if realised, including the identification of dose-response relationships, where appropriate
2. How might the receptors become exposed to the hazards and what is the probability and scale of exposure?	**Exposure assessment**: Evaluating the plausibility of the hazard being realized at the target, and by which mechanisms, allowing an assessment of the probability, magnitude and duration of exposure
3. Given exposure occurs at the above probability and magnitude, what is the probability and scale of harm?	**Risk estimation**: Consideration of the consequences of exposure with reference to effects and dose, expressed as a likelihood or probability of the hazardous effects of exposure being realized; and expressed over a range of spatial and temporal fields
4. How significant is the risk and what are the uncertainties?	**Risk characterization**: Evaluating the acceptability and significance of risk with reference to standards, targets, background risks, cost−benefit criteria or risk 'acceptability' and 'tolerability' criteria, and commenting on the uncertainties associated with the assessment

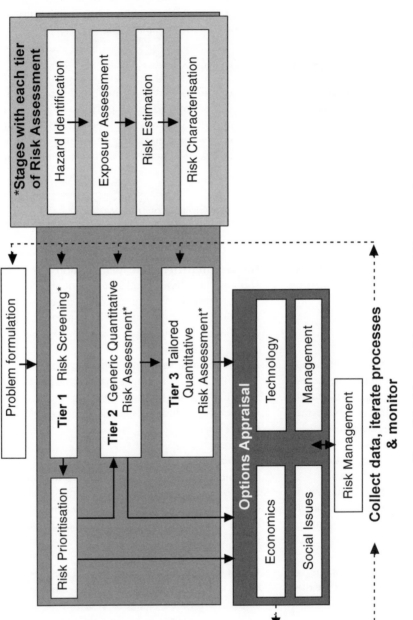

Figure 5.1   A framework for environmental risk assessment.

As a management tool, risk assessment can be conducted at various levels or tiers (Pollard *et al.* 1995, Environment Agency 1997). This can start with an initial screening of risk using a 'source-pathway-receptor' approach (Figure 5.1), and can increase in complexity through the use of risk ranking systems to a detailed analysis of complex risks using quantitative techniques to assess and express consequence and probability in numerical terms. Risk assessment is iterative and informs the appraisal of risk management options within a wider decision-making context.

Exposure assessment (Box 5.1) is an important stage of the risk-assessment process, as risks cannot occur without the exposure of a target or receptor to the source of the hazard. For chemical exposure, this principle is encapsulated within the phrase 'the dose makes the poison'; that is, it is the amount of a hazardous substance (either as a one-off dose or as incremental doses over time) that reaches a receptor that is important in determining the risk, along with the dose–response function. This reference to exposure provides us with a broader context for a discussion of pollutant transport processes, in that these processes inform exposure assessment — they represent the pathway(s) between the source of the hazard and the receptor and their characteristics determine the amount of pollutant to which the receptor is exposed.

5.3.3 FORECASTING THE FUTURE STATE OF THE ENVIRONMENT

Risk assessment has become an important tool in moving environmental regulation from a reactive to a more proactive activity. This is because risk can also be concerned with what may happen at some point in the future, should exposure to a particular hazard result in an anticipated exposure. Estimating, or forecasting what may happen as opposed to making decisions about the future based solely on the current state of the environment is, by definition more defensible. It may also prove to be more economical. Notwithstanding the above, forecasting the future state of our environment, through an understanding of the risks that may occur, is a complex area. The environment is a dynamic system, and about which comparatively little is known. As a consequence, accurate forecasts of the future state of the environment will not be possible. However, it is clear that, by looking for possible future scenarios, such as the need to regenerate brownfield sites for future house building, regulation and environmental protection effort can be focused appropriately.

Given our knowledge of how industry and other pressures such as the disposal of waste, private transport and energy usage will change in the future, we can make reasonable assumptions about the future. In addition, in some areas we can predict how the environment may respond to such changes in inputs. A considerable constraint, however, is the paucity of risk-based environmental standards. The Royal Commission on Environmental Pollution has recently expressed concern in this area (Royal Commission on Environmental Pollution 1998). In the absence of standards with an acceptable probability component, regulatory action can only be taken on when damage occurs, as opposed to when the likelihood of damage exceeds a predetermined acceptable level. The paucity of such standards will constrain our ability to forecast the severity of future environmental damage.

The needs of future generations are a particular consideration when it comes to long-term environmental management and protection. Increasingly, it will be possible to identify risks to future generations and the environment in which they may live. However, placing a value on that environment, in order to balance the costs of risk mitigation today is much more difficult, and will require some important assumptions. While our knowledge of pollutants, transport processes and their short-term effects has increased, gaps in our longer-term understanding and of the socio-economic implications of the effects warrant attention.

5.4 POLLUTANT TRANSPORT PROCESSES AND EXPOSURE ASSESSMENT IN A REGULATORY CONTEXT

5.4.1 CONTEXT

The Environment Agency of England and Wales was established on 1 April 1996. It has statutory responsibilities for water resources, pollution prevention and control, flood defence, fisheries, conservation, navigation and recreation across England and Wales. For the Environment Agency, environmental exposure assessment (Pollard *et al.* 1999) is concerned with evaluating the mechanisms, probability, duration and magnitude of exposure to a hazard. It is principally concerned with the 'pathways' by which exposure may take place. Hazard sources of interest may be **physical** (e.g. flood waters), **chemical** (e.g. oestrogenic substances) or **biological** (e.g. algal blooms) in nature, and the receptors of interest may include humankind, buildings and infrastructure, ecosystems and environmental capital (air quality, aquifers).

The study boundaries for environmental exposure assessment are potentially from the initiation of the hazard (whether intentional or accidental, natural or anthropogenic) to the point where harm is manifest and, therefore, may include:

1. Direct release from the source term to the recipient environmental medium (e.g. stack emissions to the atmospheric environment).
2. Advective transport and distribution within a recipient medium to an exposure point (e.g. transport of a point-source discharge via surface water flow).
3. Multimedia fate processes between recipient media and other indirect media, for example, the partitioning of organic compounds between recipient soil and groundwater or soil and plants (Wilson *et al.* 1996, Duarte-Davidson and Jones 1996).
4. Transport mechanisms to points of exposure (advection, diffusion and dispersion of landfill leachate or groundwater pollutants to public supply boreholes).
5. 'Dose' estimation (via direct and indirect pathways), with reference to factors influencing exposure to the hazard, ultimately as the 'effective' dose.

In practice, the detail of dose estimation, beyond consideration of an exposure point concentration (for example as an actual or predicted environmental concentration), is encapsulated during the derivation of an environmental quality standard.

5.4.2 CURRENT PRACTICE IN THE APPLICATION OF POLLUTANT TRANSPORT MODELS

Most pollutant concentrations can be estimated through monitoring or modelling. At present, environmental modelling effort focuses on the use of **environmental distribution models** concerned with the transport, degradation or multiphase partitioning of contaminants in the environment in space and time. In practice, **dose determination modelling** for the evaluation of exposure intakes is not widely applied with the exception of product licensing and radioactive waste disposal. In the future, however, this will become a wider feature of the Agency's new duties and powers on contaminated land, for example. **Pharmacokinetic modelling** for the determination of effective doses at the target organ from exposure point 'intakes' is not in routine practical use at present.

Increasingly, environmental distribution models can incorporate contaminant removal processes such as biotransformation, decay or hydrolysis; processes themselves that might be incorporated within bulk transport models. For rivers, groundwater and air, finite element dispersion modelling is used to determine concentration distributions within a medium, usually as a function of space and/or time (Douben 1998). Many models of pollution dispersion in environmental media are based on simplified parameterizations of ensemble-average conditions. This is particularly true of air dispersion models, for example, which for many years have focused on ensemble-mean dispersion rather than on trying to include the detailed (in time and space) fluctuations occurring around the ensemble mean on specific occasions (i.e. realizations). The advent of shorter-term and higher-percentile air quality standards has made it necessary to develop models that can predict the probability of such fluctuations. Generic multiphase partitioning models (e.g. USES and EUSES) are applied in chemical product licensing for predicting equilibrium distributions of contaminants across air, water, soil and biota and invoke Mackay fugacity modelling for these purposes (van Leeuwen 1995).

5.5 CASE STUDIES FROM THE GEOSPHERE

For the modelling of pollutant transport processes and the execution of exposure assessment, the geosphere presents the particular challenges of:

- extreme system heterogeneity
- multiple phases (pore fluids, mineral and organic matter, biota and often 'free product') in intimate contact and

- analytical difficulties that often hinder model validation (Hrudey and Pollard 1993).

All models in this context become a gross simplification of the field situation. Nevertheless, transport processes must sometimes be simplified for the purposes of informing regulatory decisions on the management of risk. Three case studies of regulatory exposure assessment are presented below that illustrate the role of pollutant transport processes in this regard.

5.5.1 SOIL VAPOUR INTRUSION INTO BUILDINGS FROM CONTAMINATED LAND

The increased use of brownfield sites for housing, and the concerns expressed regarding the movement of pollutants from contaminated sites into adjacent buildings has emphasized the need for the rigorous assessment of human health risks from soil pollutants. In deriving guideline values to protect against unacceptable risks to human health from land contamination, the Department of the Environment, Transport and the Regions has developed an exposure assessment model (Contaminated Land and Exposure Assessment (CLEA) model) that incorporates a number of exposure routes. For volatile substances,

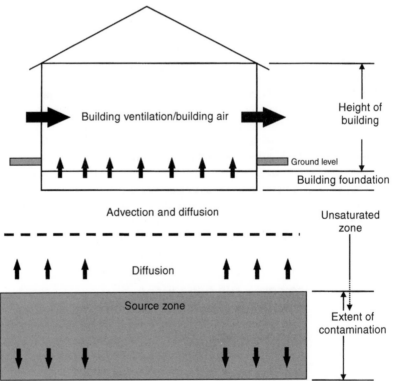

Figure 5.2   Soil vapour intrusion into buildings from contaminated land.

soil vapour intrusion from surface or subsurface contamination in the geosphere can be an important route of exposure.

A wide variety of models have been developed to predict soil vapour concentrations based on pollutant concentrations measured in soils, building design and the most likely migration routes (see for example Johnson and Ettinger 1991, Ferguson *et al*. 1995, Hers *et al*. 1997, Johnson *et al*. 1998). Figure 5.2 is an illustrative example of such models and shows a one-dimensional steady-state transport model based on a multi-compartment framework is presented, where the relevant compartments are the source of the pollution, the unsaturated soil zone and the building foundation. Models like this can be used to estimate contaminant levels inside buildings from the soil vapour concentration based on the equilibrium partitioning between the absorbed, aqueous and vapour phases, and are based on octanol–water partition coefficient, organic carbon content and effective diffusivity.

The complexity in the estimates produced by individual models can vary considerably, reflecting the purpose for which the model is being developed. Generally, models incorporate the fate and transport processes in the unsaturated zone and transport through the building foundation, taking into account the foundation construction, building ventilation rates, enclosure height and estimated flux from subsurface sources. More complex models may consider degradation rates and fate and transport processes in the source zone.

5.5.2 NEW SUBSTANCES REGULATION

European Union (EU) legislation requires that, to comply with the 'new substances' regulations, the appropriate authorities, including the Environment Agency, need to collect relevant documentation to assess the intrinsic hazardous properties of substances prior to agreements on product licensing. Regulatory authorities conduct a preliminary evaluation of human health risks, using the limited information normally supplied regarding exposure. All risk assessments of new chemicals must be conducted in accordance with guidelines set out in the European Commission's *Technical Guidance Document (TGD) on Risk Assessment of New and Existing Chemicals* (European Commission 1996). The guidance documents are used by the Environment Agency to assess the intrinsic hazardous properties of these substances before, for example, introduction into the market. The *TGD* document provides guidance on the assessment of risks from chemicals to humans and the environment and uses simple compartmental models for surface water, groundwater, soil and air and a multimedia approach for estimating exposure to humans.

Exposure assessment for the soil compartment is important with respect to terrestrial organisms. Contaminated soil may also lead to indirect human exposure via the consumption of crops grown on agricultural soils or cattle grazing on contaminated grassland. Figure 5.3 shows the key fate-determining processes of chemicals in soil. When considering exposure of contaminants in

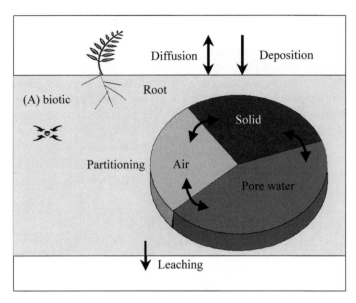

Figure 5.3 Possible fate processes of chemicals in the soil compartment (adapted from van de Meent *et al.* 1995).

the soil environment, the *TGD* considers the exposure from sludge-amended soils and from dry and wet deposition from the atmosphere. The mobility of chemicals in the soil is largely determined by the partitioning between the solids, air and pore water in soil, which can be estimated from the adsorption/desorption rate constants and the Henry's law constant. Soil sorption influences migration of chemicals down the soil profile and volatilization from the soil surface. In addition, the bioavailability to, and biotransformation by soil micro-organisms is dependent on the fraction of the chemical not adsorbed to the solid fraction of the soil.

5.5.3 PROTECTION OF WATER RESOURCES FROM SOIL-BORNE POLLUTANTS

The Environment Agency's Groundwater Protection Policy (for context and explanation, see United Kingdom Groundwater Forum 1998) sets out a framework of guidance to ensure that groundwater resources are safeguarded from pollution. A serious risk to groundwater quality can arise from accidental spills or leaks from tanks and pipelines of petroleum products and solvents. Where only sparingly soluble, the bulk or 'free products' (that is, liquid solvents), are referred to as non-aqueous phase liquids (or NAPLs). They can be further subdivided into light (L) and dense (D) NAPLs, according to whether they are lighter or denser than water. The presence of NAPLs in the unsaturated and saturated zones of subsurface soils offers a potential source of ongoing pollution to ground and surface waters. Because of the nature of these compounds, they present a formidable multiphase problem with respect to their remediation (Figure 5.4). Following a spillage, LNAPLs will percolate through the unsaturated zone of the soil. Once they reach the surface of the groundwater table, they form a floating pool of material; the soluble constituents then dissolving into the groundwater and migrating in the direction of groundwater flow. If a significant portion of the LNAPL is retained in the unsaturated zone, volatilization can become an important pathway. Dense DNAPLs are transported by gravity through the unsaturated zone, and once it reaches the saturated zone a plume can develop of dissolved chemical. Ultimately, the sinking free product can descend to the bedrock, forming a pool of free product that is almost unrecoverable.

The derivation of remediation objectives for soils posing a threat to water resources relies on a back-extrapolation of water quality standards at the receptor to the soil or groundwater concentration at the source: an exposure assessment operating in 'reverse'. A tiered approach (Figure 5.5) has been developed by the Environment Agency that involves structured decision-making, cost–benefit analysis and progressive data collection and analysis (Environment Agency 1999). At each tier, a remediation objective is derived that can form the starting-point for regulatory decisions on the standard of remediation. With successive tiers the data requirements and sophistication of the analysis increases, with a corresponding reduction in the uncertainty of the predicted impact (Figure 5.5). The increase potentially allows for a relaxation in the remedial requirements, if the risk assessment is favourable. Tier 1 considers whether the concentration in pore water in polluted soil below the site is sufficient to impact on the receptor, ignoring dilution, dispersion and attenuation processes. Tier 2 takes into account the dilution effect in the receiving groundwater or surface water body and whether this is sufficient to reduce the pollutant concentration to acceptable levels. Tiers 3 and 4 consider whether natural attenuation of the pollutant, as it moves through the unsaturated and saturated zones to the receptor, are sufficient to reduce contaminant concentrations to acceptable levels. The difference between Tiers 3 and 4 is that Tier 4 uses more sophisticated numerical models and predictive processes to calculate the significance of the attenuation. For each tier, the pore water concentration determined for the soil zone is compared to a remedial objective to determine the need for remedial action.

These case studies illustrate the link between pollutant fate and transport, environmental forecasting in the context of predictive modelling and environmental exposure assessment within the broader context of risk assessment. After the uncertainties associated with the baseline toxicology, it is often the exposure assessment that harbours the greatest source of uncertainty for risk assessors. From a regulatory perspective, decisions must be based on sound science, but they will also often be made in the face of considerable conceptual and data uncertainty. This dilemma can be addressed in part by risk assessors ensuring:

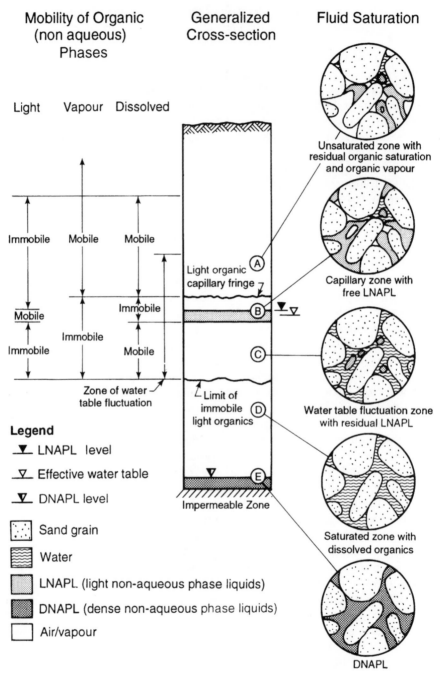

Figure 5.4 Multiphase pollution of non-aqueous phase liquids in the subsurface (after Hrudey and Pollard, 1993).

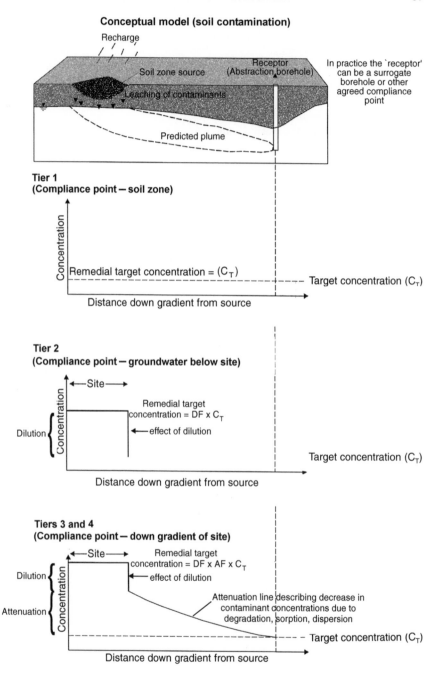

Figure 5.5 Summary of tiered approach for the derivation of remedial objectives to protect water resources from land contamination (Environment Agency 1999).

1.   The conceptual model of exposure is bounded and agreed by all parties early on.
2.   A clear, unambiguous guiding rationale to the exposure assessment is presented prior to detailed analysis.
3.   Uncertainties in the exposure assessment are recognized and prioritized.
4.   Input data is of a high quality and validated.
5.   The level of analysis in the exposure assessment is proportionate to the problem and input data quality.
6.   Assumptions are clearly justified and referenced.

5.6 CONCLUSIONS

1.   Forty years of research in pollutant transport processes has enlightened the environmental science and engineering community as to the fate and behaviour of pollutants in the environment.
2.   Many of these processes have been modelled for use in compliance assessments in regulation, starting first with the aquatic and air environments where homogeneity has often been assumed, but moving beyond to the geosphere.
3.   The development of environmental risk assessment as a management tool for assessing the risks of pollutants to humans and the environment embraces pollutant transport as a component of the exposure assessment.
4.   Environmental legislation and regulatory guidance is now actively adopting the outputs of the research community to better inform regulatory decision-making.

ACKNOWLEDGEMENTS

The authors wish to acknowledge colleagues in the Groundwater and Contaminated Land National Centre of the Environment Agency for assistance with the preparation of figures.

REFERENCES

Alloway BJ and Ayers DC (1993) *Chemical Principles of Environmental Pollution.* Blackie Academic and Professional, London.
Calow P (1998) *Handbook of Environmental Risk Assessment and Management.* Blackwell Science, Oxford.
Department of the Environment (1995) *Risk Assessment and Risk Management for Environmental Protection.* Her Majesty's Stationery Office, London (under revision).
Douben PET (1998) *Pollution Risk Assessment and Management.* Wiley, Chichester.
Duarte-Davidson R and Jones KC (1996) Screening the environmental fate of organic contaminants in sewage sludge applied to agricultural soils: II. The potential for transfer to plants and grazing animals. *Science of the Total Environment*, **185**, 59–70.
Environment Agency (1997) *A Guide to Risk Analysis at the National Centre for Risk Analysis and Options Appraisal.* Environment Agency, London.
Environment Agency (1999) *Integrated Methodology for the Derivation of Remedial Targets for Soil and Groundwater to Protect Water Resources.* R&D Publication 20, Environment Agency, Bristol.

European Commission (1996) *Technical Guidance Document in Support of Commission Directive 93/67/EEC on Risk Assessment for New Notified Substances and Commission Regulation (EC) No. 1488/94 on Risk Assessment for Existing Substances. Part I*. Office for Official Publication of the European Commission, Luxembourg.

European Environment Agency (1998) *Environmental Risk Assessment: Approaches, Experiences and Information Sources*. Office for Official Publications of the European Communities, Luxembourg.

Ferguson CC, Krylov VV and McGrath PT (1995) Contamination of indoor air by toxic soil vapours: a screening risk assessment model. *Building And Environment*, **30**, 375–385.

Hers I, Zapf-Gilje R, Petrovic R, Macfarlane M and McLenehan R (1997) Prediction of risk-based screening levels for infiltration of volatile sub-surface contaminants in buildings. In *Environmental Toxicology and Risk Assessment (Sixth Volume)*, Dwyer FJ, Doane T and Hinman ML (eds), American Society for Testing and Materials.

Hester RE and Harrison RM (1998) Risk assessment and risk management. *Issues in Environmental Science and Technology*, **9**, 1–168.

Hrudey SE and Pollard SJT (1993) The challenge of contaminated sites: remediation approaches in North America. *Environmental Reviews*, **1**, 55–72.

Johnson PC and Ettinger R (1991) Heuristic model for predicting the intrusion rate of contaminant vapours into buildings. *Environmental Science and Technology*, **25**, 1445–1452.

Johnson PC, Kemblowski MW and Johnson RL (1998) *Assessing the Significance of Subsurface Contaminant Vapour Migration to Enclosed Spaces—Site Specific Alternatives to Generic Estimates*. American Petroleum Institute. Health and Environmental Sciences Department, Publication Number 4674.

Mackay D (1991) *Multimedia Environmental Models: the Fugacity Approach*. Lewis Publishers, Chelsea.

Masters GM (1991) *Introduction to Environmental Engineering and Science*. Prentice-Hall, New Jersey.

Pollard SJT, Harrop DO, Crowcroft P, Mallett SH, Jefferies SR and Young PJ (1995) Risk assessment for environmental management: approaches and applications. *Journal of the Chartered Institution of Water and Environmental Management*, **9**, 621–628.

Pollard SJT, Timmis R and Robertson S (1999) Environmental exposure assessment in the Environment Agency. In *Exposure Assessment in the Evaluation of Risk to Human Health*, RATSC Workshop Report (cr5) Institute for Environment and Health, Leicester, pp. 17–21.

Royal Commission on Environmental Pollution (1998) *Twenty-First Report: Setting Environmental Standards*. The Stationery Office, London.

Schnoor JL (1996) *Environmental Modeling: Fate and Transport of Pollutants in Water, Air and Soil*. Wiley Interscience, Chichester.

United Kingdom Groundwater Forum (1998) *Groundwater: Our Hidden Asset*. British Geological Survey, Nottingham.

van de Meent D, de Bruijn JHM, de Leeuw FAAM, de Nijs ACM, Jager DT and Vermeire TG (1995) Exposure modelling. In *Risk Assessment of Chemicals: An Introduction*, van Leeuwen CJ and Hermens JLM (eds), Kluwer, Dordrecht, pp. 103–145.

van Leeuwen CJ (1995) General introduction. In *Risk Assessment of Chemicals: An Introduction*, van Leeuwen CJ and Hermens JLM (eds), Kluwer, Dordrecht, pp. 1–17.

Wilson SC, Duarte-Davidson R and Jones KC (1996) Screening the environmental fate of organic contaminants in sewage sludge applied to agricultural soils: I. The potential for downward movement to groundwaters. *Science of the Total Environment*, **185**, 45–57.

6

The EU-TGD for New and Existing Substances: Does it Predict Risk?

TJALLING JAGER AND JACK H. M. DE BRUIJN

*National Institute of Public Health and the Environment (RIVM),
Bilthoven, The Netherlands*

6.1 INTRODUCTION

Chemical risk assessment tries to protect humans and the environment from the possible adverse effects of substances. In the European Union, chemical risk assessment has become a very important part of the management of new industrial chemicals that are placed on the market as well as for existing substances. During the last few years legislation has come into force which describes basic principles for the risk assessment of chemicals. For new chemicals these principles have been laid down in Commission Directive 93/67/EC (EC 1993) and for existing chemicals in Commission Regulation No. 1488/94 (EC 1994). Since this legislation is restricted to the definition of terminology and a description of the basic elements of the risk assessment process, there was a pressing need for more detailed guidance on how actually to perform the risk assessments in practice. Therefore, from 1992 to 1996 the European Commission in collaboration with the Member States and the European chemical industry worked together on the development and harmonization of risk-assessment methodologies. Finally, this led to the publication of what is now called the '*TGD*': the *Technical Guidance Document on Risk Assessment for New Notified and Existing Substances* (EC 1996b).

The methodology described in the *TGD* follows the basic framework of the risk-assessment process as given in Figure 6.1. Key parts of this process are, on the one hand, the exposure assessment resulting in a so-called predicted environmental concentration (PEC) using different kinds of data and models on emissions and distribution in the environment and, on the other hand, the effects assessment where, on the basis of different kinds of ecotoxicity data, a so-called predicted no-effect concentration (PNEC) for the environment is calculated. In the risk characterisation phase, PEC and PNEC values are compared in order to decide whether or not the risks are acceptable or whether further information or testing is needed to refine the risk quotients. These PEC/PNEC ratios are calculated for all environmental protection targets as defined by the *TGD*: aquatic ecosystems,

Forecasting the Environmental Fate and Effects of Chemicals. Edited by Philip S. Rainbow, Steve P. Hopkin and Mark Crane.
© 2001 John Wiley & Sons Ltd

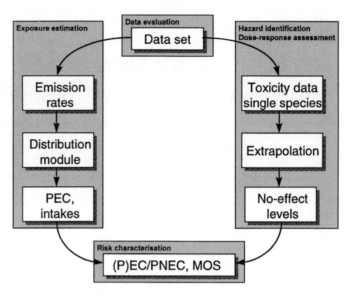

Figure 6.1   The basic framework of the risk assessment process.

terrestrial ecosystems, sediment ecosystems, top predators and micro-organisms in sewage-treatment systems. To ensure rapid and transparent assessments, the methodology of the *TGD* has been implemented in a computerized system: the European Union System for the Evaluation of Substances (EUSES) (EC 1996a, Vermeire *et al.* 1997). With this system calculations can be performed for, in principle, all chemicals, based on a relatively limited amount of available data (e.g. the EC base set for new chemicals).

6.2   DOES THE TGD PREDICT RISKS?

It is important that the risk assessor who follows the procedures and models as described in the *TGD* and implemented in EUSES is aware of the 'validity' of this system when evaluating the outcome. By understanding the validity, the user will be able to make a better judgement of the possibility that the predicted risks of the use of a certain substance to the protection targets may also occur in reality. It is, however, difficult to specify the degree of certainty that a decision-maker needs when assessing these risks. Furthermore, the degree of certainty depends heavily on the amount and quality of the input data: no system may be expected to provide accurate estimates of exposure and effects on the basis of base-set data alone. Nevertheless, the user of a system should be aware of the degree of (in)accuracy of the models so that this information can be taken into account (quantitatively or qualitatively) in the decision-making process. Therefore, the principal aim for validation of these types of systems should be to show transparently how well the model represents a part of reality (Jager 1995). It is up to the decision-maker to judge whether or not this accuracy is sufficient to justify risk reduction measures. Furthermore, it is important to indicate for which part of reality the model provides adequate results (Schwartz 1997).

A strict validation of the risk assessment methodology as given in the *TGD* is not possible. Firstly, as discussed above, because criteria for validity are not clearly defined and, secondly, because the results of the risk characterization phase cannot be determined in the real world. The risk estimates are PEC/PNEC quotients which, strictly speaking, are not risk estimates. These ratios do not quantify the incidence and severity of toxic effects and are thus merely surrogate indicators for the unknown risk. Despite these considerations, an evaluation in a less strict manner should be performed to clarify the degree of confidence in the final results. Parts of the methodology (modules or models) can be validated numerically. Exposure concentrations can be measured, but one has to realize that measured data are usually not representative for the situation described by the TGD for several reasons:

1. The assessment uses the concept of a standard non-existent model environment with predefined characteristics. These characteristics can be average values or reasonable worst-case values depending on the parameter in question. Measured field data will invariably be non-representative for the model situation. The concept of a standard scenario clearly plays a crucial role in the assessment and its applicability and appropriateness should be considered in a model validation.

2. Since it must be possible to perform the assessment with a limited data set (e.g. the base set for new chemicals), several chemical-specific parameters are set to reasonable worst-case values in the absence of specific data (e.g. release rates, degradation rates).

3. Most variations in time and space are averaged out in the local and regional models described in the *TGD*. It is not abnormal for the concentration of a chemical in sediment samples from within the same square metre to vary by more than a factor of 10 or even 100 (ECETOC 1992). This variation cannot be taken into account in the *TGD* methodology.

The use of a standard scenario does not mean that the results obtained according to the *TGD* models are 'not valid'. Instead, one has to take care that measured data are only used in a comparison when they are representative of the situation that the *TGD* is trying to model (compare like with like). If there is a need to predict site-specific concentrations taking into account the local characteristics, the *TGD* models can be used as a starting-point, but better models are available to obtain these estimates, at the cost of much greater data requirements.

6.3 VALIDATION STATUS OF TGD MODULES

In this section a few modules of the *TGD* are discussed as examples to clarify the validation status of the whole risk assessment methodology. When possible, a brief comparison between model results and measured data is presented. A more thorough discussion of these and other modules is given in Jager (1995).

6.3.1 EMISSION

When no specific information on chemical release is available standard tables are used, based on the use pattern of the chemical. The values in these tables are partly based on in-depth studies but also on professional judgement. Emission factors concern the size (capacity) of the main source and the activity level (number of emission days). In generating these factors the aim was to obtain estimates which were within an acceptable range of the expected level. For estimates based on, for example, use category documents, better values may be expected with an estimated accuracy within a factor of two. Unfortunately, current experience with EU work on existing chemicals has shown that, except for production, specific release information is very difficult to obtain. As the more specific use category documents are only available for a limited number of industrial categories, the exposure assessment is carried out with the *TGD*-default emission factors in most cases. To quantify this uncertainty, the present EU risk assessment reports on existing chemicals were screened for the availability of site-specific release estimates. These specific release estimates were then compared with the *TGD* defaults for the corresponding life cycle stage. For 20 chemicals, site-specific release factors could be found for various life cycle stages. The correlation between site-specific emission factors versus the defaults from the A-tables in the *TGD* for the 20 compounds is shown in Figure 6.2. The scatter plot comprises the data (water and air) for all available life

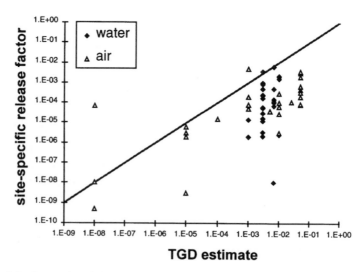

Figure 6.2   Scatterplot of site-specific emission factors versus *TGD* defaults (production and processing). The 1 : 1 line is shown.

cycle stages. The general conclusion from this limited survey is that the *TGD* defaults are generally higher, i.e. more conservative, than the site-specific release factors. In some cases the difference between site-specific and default is up to three, in extreme cases up to four, orders of magnitude. In other cases, the estimate is quite accurate. The current *TGD* defaults can thus be considered as precautionary estimates which are in many, but not all, cases quite conservative.

6.3.2 SEWAGE TREATMENT

The multi-media model SimpleTreat 3.1 (Struijs 1996) is mentioned in the *TGD* as a possible model to estimate the fate of substances in a sewage-treatment plant (STP). This multimedia box model of the 'Mackay type' has been implemented in EUSES. The modelled STP is a municipal STP with standard characteristics, although many industrial plants have their own dedicated STPs which are usually considerably larger and more efficient. The previous version of SimpleTreat (version 1.0) was investigated earlier (Struijs *et al.* 1991, Toet *et al.* 1991). For a wide range of substances and situations, the model reasonably predicts effluent concentrations and sludge concentrations (generally within a factor of 10) with the standard scenario. When the parameters are adapted to the operating parameters of a specific plant, the model is able to simulate field data within a factor of 1.5. An extensive validation study for STP models was recently conducted using a pilot-scale activated sludge plant (Temmink and Klapwijk 1998). It was concluded that existing models are applicable for a screening stage assessment. Biodegradation kinetics are the most important bottleneck in the application of models as linear kinetics are insufficient. It was shown that an increasing loading rate of a compound is counteracted by an increased biodegradation capacity of the activated sludge. This indicates Monod kinetics rather than a first-order rate constant which is applied in the *TGD* as a default for chemicals that are positive in OECD biodegradability tests. It seemed that the Monod parameters are not constants but increase at decreasing influent chemical concentrations. However, the validation study was limited to only a few easily biodegradable substances and further work needs to be done.

6.3.3 DEGRADATION RATES

Unless kinetic data from simulation tests are available, first-order rate constants are based on the results of standard biodegradability tests (OECD 1993). These 'ready' or 'inherent' tests are conducted in artificial systems and no attempts are made to reflect the real world. Rate constants from real-world studies are scarce, however, and usually there is no alternative to applying results of those artificial standard tests. Because of the low predictive power of these tests, worst-case rates are assigned to results which have been categorized based on expert judgement (VROM 1995). This extrapolation procedure represents a consensus, yielding rather conservative rate constants as defaults for aerobic biodegradation in water

Table 6.1   Rate constant for half-life in surface water based on the results of screening tests on biodegradability

Result from standard test	Half-life (days)
Readily biodegradable	15
Readily biodegradable, but failing 10-day window	50
Inherently biodegradable	150
Not biodegradable	Infinite

and soil. The extrapolation for surface water is shown in Table 6.1 to illustrate this procedure.

Thirty-six substances were selected because results from screening, soil and surface-water biodegradation tests were available for them. Figure 6.3 compares measured half-lifes reported (ECETOC 1994, Struijs and Van den Berg 1994) with half-lives obtained from OECD standard tests results, categorised according to the TGD. For the soil compartment, the comparison is based only on data from an ECETOC (European Centre for Ecotoxicology and Toxicology of Chemicals) database (see Figure 6.4). Real world biodegradation rates may differ from rates obtained according to the *TGD* approach by several orders of magnitude. The *TGD* is generally on the conservative side (up to a factor of 100) but, in a few cases, the measured degradation rates are slower than the estimates. For the sediment compartment, such a comparison is not yet possible due to the lack of field data on aerobic biodegradation. Anaerobic biodegradation, especially reductive dehalogenation, has been reported for a reasonable number of compounds. However, data based on a standardized biodegradation test for anaerobic biodegradation are the limiting factor in developing an appropriate extrapolation routine.

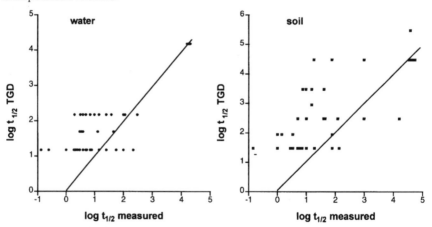

Figure 6.3   Measured versus evaluated (according to the *TGD*) logarithmic half-lives in surface water and soil. Data of ECETOC (1994) and Struijs and van den Berg (1994) are combined. The 1 : 1 line is shown.

6.3.4 REGIONAL DISTRIBUTION

For regional and continental distribution, a multimedia box model of the 'Mackay type' is used in EUSES (a simplified version of SimpleBox 2.0; Brandes *et al.* 1996). Region and continent are made up of six well-mixed compartments (air, water, sediment, and three soil compartments). The region comprises $200 \times 200\,km$ with averaged EU properties; the continent has the size of all EU Member States plus Norway. All of the emissions into these compartments are treated as continuous and diffuse which allows steady-state concentrations to be calculated. These are used in the comparison with PNECs although, for some chemicals, it may take several hundreds or thousands of years to achieve a steady state.

The main problem with validation of this regional distribution model is the lack of good-quality data. The available data are usually not representative for the generic SimpleBox environment and the model predicts long-term average concentrations which cannot usually be determined in the field. Some monitoring data were taken from EU risk assessment reports and compared to PEC estimates on regional and continental scales according to the *TGD* models. It is difficult to draw definite conclusions on this limited verification, but it may provide an impression of the order of magnitude. The SimpleBox estimate is usually at the lower end or below the reported values (Figure 6.4). This is to be expected as measurements will usually be conducted when and where a problem is expected. The maximum reported values can be much higher (up to a factor of 10^7!) which is to be expected as these were usually point-source related (and thus not comparable to the regional estimate). The modelled air concentration seems most representative for the measured values, which is not surprising as this compartment is better described by 'well-mixed' than the other media. Soil concentrations are reported for one chemical only. Sediment and groundwater seem usually to be underestimated.

Owing to the assumptions, the Simplebox model as implemented in EUSES is not easily validated, and the parameter values for the average EU environment need reconsideration. It should be investigated whether another definition of the standard region will affect the model results in a significant manner (e.g. a Scandinavian or southern-Europe scenario). Nevertheless, the limited data suggest that the user should be aware that SimpleBox will usually give an estimate at the lower end of the range.

6.3.5 ECOLOGICAL EFFECTS ASSESSMENT

According to the *TGD*, predicted no-effect concentrations (PNEC) for ecological endpoints are derived by applying fixed assessment factors to the results of single-species laboratory tests. In using the assessment factors, it is assumed that laboratory animals are representative of field species and that protection of the most sensitive species will also protect ecosystem structure and function. Power and McCarty (1997) critically discussed these basic assumptions and concluded

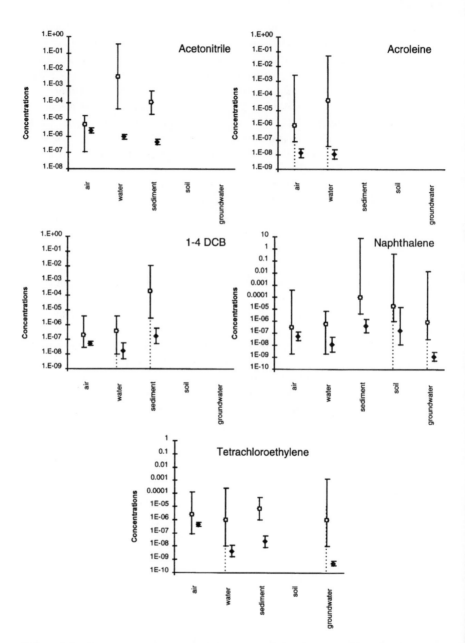

Figure 6.4 Comparison of measured (open squares) and estimated (closed symbols) environmental concentrations). The measured data point is a 'typical value', the range represents the minimum reported value (a dotted line means lower than detection limit) and the maximum value related to a point source.

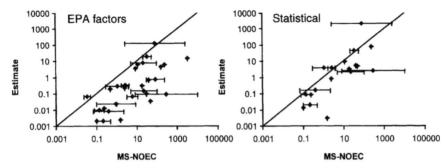

Figure 6.5 Comparison of estimated PNECs with NOECs from multiple-species (mesocosm) experiments (MS-NOEC). Two estimation approaches are shown: EPA assessment factors and the statistical extrapolation according to Aldenberg and Slob (1993).

Table 6.2 PNEC values (μg l^{-1}), expressed as dissolved concentrations) for LAS, AE, AES and soap based on single species data and field NOECs

Surfactant	PNEC based on single-species data	Range of field NOECs	Final PNEC
LAS C$_{11.6}$	320	250–500	250
AE C$_{13.3}$ EO$_{8.2}$	110	42–380	110
AES C$_{12.5}$ EO$_{3.4}$	400	190–3700	400
Soap	27	—	27

that they may be useful from a policy perspective but are not scientifically valid. However, alternatives are not available. Emans *et al.* (1993) compared data from multiple-species experiments with results from different extrapolation methods on single-species tests. The best estimates of 'safe levels' were derived using the statistical extrapolation method of Aldenberg and Slob (1993). The EPA assessment factors were more conservative, but usually safe enough. The data are shown in Figure 6.5.

In a recently performed risk assessment of four high volume surfactants carried out in the Netherlands, a comparison was made between the PNEC values obtained with the statistical extrapolation technique as developed by Aldenberg and Slob (1993) and NOEC values obtained in a relatively high number of (semi-)field tests. The results of this study are summarised in Table 6.2. From these results it was concluded that there is a remarkable agreement between the extrapolated PNEC and the lowest NOECs from the field tests (van de Plassche *et al.* 1999).

Soil and sediment PNECs are often derived through equilibrium partitioning since specific data for these compartments may be lacking. With this method it is assumed that soil and sediment organisms are as sensitive as water organisms and are exposed through pore water. At log $Kow > 5$ the *TGD* calls for an extra assessment factor of 10 to account for additional uptake through ingestion of contaminated particles. Belfroid *et al.* (1996) reviewed the equilibrium partition approach and concluded that the main route of uptake of hydrophobic chemicals in invertebrates from soil and sediment is through interstitial water. For compounds with a log $Kow > 5$, uptake from particles becomes important, although the quantitative impacts are not clear.

Based on the limited data, the assessment factor approach seems to provide a safe, but worst-case, estimate. It is possible to derive data-based assessment factors or use statistical approaches, but these methods are not yet acceptable to all Member States (Jager 1998a,b).

6.4 CONCLUSIONS

Environmental risk assessments performed according to the methodology as described in the *TGD* provide the risk assessor with quantitative estimates of the potential exposure and predicted no-effect concentrations for the different protection targets. In addition, a simple quantitative method is given whereby PEC and PNEC are compared and the results are used to make a decision on the acceptability of the risks of substance use. Depending on the type of data available, an indication can be obtained of whether these risks will actually occur, and if so, which protection targets are most likely to be harmed. However, because of the nature of the system and the concept of a 'standard environment' one cannot expect the risk assessment methodology as described in the *TGD* to predict spatially and temporally varying concentrations in the environment with a great degree of accuracy. In a preliminary assessment the model's outcome will be a conservative estimate that supports decision-making on chemical safety based on the legally restricted minimum requirements of the base set data. Substitution of defaults with more appropriate (post base set) data will automatically increase the accuracy and realism of the predictions and the *TGD* allows the risk assessor to do this.

Currently, the validation status of the *TGD* methodology as a whole leaves much to be desired. The limited validation status of the *TGD* does not mean that the system should not be used in risk assessment. It represents the 'state-of-the-art' in generic risk assessment methodology, and Member States as well as the European chemical industry have reached consensus over its contents. Nevertheless, the user of the *TGD* should be aware of the degree of accuracy and precision of the specific models fully to appreciate the results. Clearly, the degree of deviation from reality may be substantial for some of the outcomes of the models. Several modules can be characterized as rather conservative: release estimation, biodegradation, the exposure scenario and the environmental effects assessments. The estimation of partition coefficients and BCFs is, however, an average case because log-linear regressions on experimental data are applied to estimate these parameters. The regional distribution model may be characterized as quite optimistic as it usually provides estimates at the lower end of the range of measured values.

It is important to realize that the *TGD* risk assessment methodology should be able to distinguish potentially risky chemicals from harmless substances. This was recognized in an earlier report (Jager 1995), but is often overlooked when validity is discussed. Owing to the generic nature of *TGD* assessment, the applied models usually have a low degree of detail. It is therefore tempting to apply more site-specific models in the assessment, although this will inevitably lead to large data needs and results with limited general applicability. Implementation of an uncertainty analysis in the risk characterization phase might be a more useful alternative in this respect. Currently, decisions on chemical safety are made on the basis of point estimates (PEC/PNEC). It is however advisable to take uncertainties in the risk assessment explicitly into account in the decision-making stage (see e.g. Jager *et al.* 2000). Furthermore, uncertainty analysis can also facilitate the way we look at 'validity'. Poor, but generally applicable, models can be useful as long as the uncertainty in them is transparently accounted for. Analysis of the main sources of uncertainty will show the risk assessor where further model or data refinement will lead to a maximum increase in the realism of the risk estimates.

ACKNOWLEDGEMENTS

The work presented here is a summary from an RIVM report (Jager 1998) and we would therefore like to express our thanks to the contributors to that report: P. Gingnagel, C. W. M. Bodar, H. A. den Hollander, P. van der Poel, M. G. J. Rikken, J. Struijs, M. P. van Veen and T. Vermeire. The reader may resort to this report for more in-depth discussion on the validation status of the TGD and EUSES.

REFERENCES

Aldenberg T and Slob W (1993) Confidence limits for hazardous concentrations based on logistically distributed NOEC toxicity data. *Ecotoxicology and Environmental Safety*, **25**, 48–63.

Belfroid AC, Sijm DTHM and van Gestel CAM (1996) Bioavailability and toxicokinetics of hydrophobic aromatic compounds in benthic and terrestrial invertebrates. *Environmental Reviews*, **4**, 276–299.

Brandes LJ, Den Hollander H and van de Meent D (1996) Simple Box 2.0: A nested multimedia fate model for evaluating the environmental fate of chemicals. Bilthoven, National Institute of Public Health and the Environment (RIVM). Report No. 719101 029.

EC (1993) Commission Directive 93/67/EEC of 20 July 1993, laying down the principles for the assessment of risks to man and the environment of substances notified in accordance with Council Directive 67/548/EEC. *Official Journal of the European Communities*, L227.

EC (1994) Commission Regulation (EC) 1488/94 of 28 June 1994, laying down the principles for the assessment of risks to man and the environment of existing substances in accordance with Council Regulation (EEC) No. 793/93. *Official Journal of the European Communities*, L161.

EC (1996a) *EUSES, the European Union System for the Evaluation of Substances*. National Institute of Public Health and the Environment (RIVM), the Netherlands. Available from the European Chemicals Bureau (EC/DGXI), Ispra, Italy.

EC (1996b) *Technical Guidance Documents in Support of Directive 93/67/EEC on Risk Assessment of New Notified Substances and Regulation (EC) No. 1488/94 on Risk Assessment of Existing Substances (Parts I, II, III and IV). EC Catalogue Numbers CR-48-96-001, 002, 003, 004-EN-C.* Office for Official Publications of the European Community, 2 rue Mercier, L-2965 Luxembourg.

ECETOC (1992) *Estimating environmental concentrations of chemicals using fate and exposure models.* Technical Report No. 50, ECETOC, Brussels, Belgium.

ECETOC (1994) *Environmental Exposure Assessment.* Technical Report No. 61, ECETOC, Brussels, Belgium.

Emans HJB, van de Plassche EJ, Canton JH, Okkerman PC and Sparenburg PM (1993) Validation of some extrapolation methods used for effect assessment. *Environmental Toxicology and Chemistry,* **12** , 2139–2154.

Jager DT (1995) *Feasibility of Validating the Uniform System for the Evaluation of Substances (USES).* Bilthoven, National Institute of Public Health and the Environment (RIVM). Report No. 679102 026.

Jager T (1998a) *Uncertainty Analysis of EUSES: Interviews with Representatives from Member States and Industry.* Bilthoven, National Institute of Public Health and the Environment (RIVM). Report No. 679102 047.

Jager T (ed.) (1998a) *Evaluation of EUSES: Inventory of Experiences and Validation Activities.* Bilthoven, National Institute of Public Health and the Environment (RIVM). Report No. 679102 048.

Jager T, Vermeire TG, Rikken MGJ and van der Poel P (2000) Opportunities for a probabilistic risk assessment of chemicals in the European Union. Accepted for publication in *Chemosphere.*

OECD (1993) Guidelines for testing of chemicals. OECD, Paris, France.

Power M and McCarty LS (1997) Fallacies in ecological risk assessment practices. *Environmental Science and Technology,* **31**, 370–375.

Schwartz S (1997) Organische Schadstoffe in der Nahrungskette. Vorstudie zur Validierung von Expositionsmodellen. Beiträge des Instituts für Umweltsystemforschung der Universität Osnabrück. Beitrag Nr. 5.

Struijs J (1996) *SimpleTreat 3.0: A Model to Predict the Distribution and Elimination of Chemicals by Sewage Treatment Plants.* Bilthoven, National Institute of Public Health and the Environment (RIVM). Report No. 719101 025.

Struijs J, Stoltenkamp J and van de Meent D (1991) A spreadsheet-based box model to predict the fate of xenobiotics in a municipal wastewater treatment plant. *Water Research,* **25**, 891–900.

Struijs J and van den Berg R (1994) Standardized biodegradability tests: Extrapolation to aerobic environments. *Water Research,* **29**, 255–262.

Temmink H and Klapwijk B (1998) *Fate of Organic Compounds in a Pilot-scale Activated Sludge Plant: DynTreat Model Validation Study.* RIZA report 98.059, Lelystad, The Netherlands.

Toet C, De Nijs ACM, Vermeire TG, van der Poel P and Tuinstra J (1991) *Risk Assessment of New Chemical Substances; System Realisation and Validation II.* Bilthoven, National Institute of Public Health and the Environment (RIVM). Report No. 679102 004.

Vermeire TG, Jager DT, Bussian B, Devillers J, Den Haan K, Hansen B, Lundberg I, Niessen H, Robertson S, Tyle H and van der Zandt PTJ (1997) European Union System for the Evaluation of Substances (EUSES). Principles and structure. *Chemosphere,* **34**, 1823–1836.

van de Plassche EJ, de Bruijn JHM, Stephenson RR, Marshall SJ, Feijtel TCJ and Belanger SE (1999) Predicted no-effect concentrations and risk characterization of four surfactants: linear alkylbenzene sulfonate, alcohol ethoxylates, alcohol ethoxylated sulfates, and soap. *Environmental Toxicology and Chemistry,* **18**, 2653–2663.

VROM (1995) Special EU expert meeting on biodegradation. *Summary Report, May 31–June 1.* Ministry of Housing, Spatial Planning and Environmental Protection, The Hague, the Netherlands.

7

Can a Substance-specific Chemical Approach Forecast the Toxicity of Effluents?

MARCEL TONKES

Department of Water Pollution Control, Institute for Inland Water Management and Waste Water Treatment (RIZA), Lelystad, The Netherlands

7.1 INTRODUCTION

In 1970 a law on the pollution of surface waters was introduced in the Netherlands. This law, the WVO, has now been in force for over 27 years. During this time, water quality regulators in the Netherlands have performed assessments of effluents by monitoring specific substances, but it has become clear that this substance-specific approach has many limitations. These limitations make it almost impossible to perform a complete assessment of effluent samples, which are often complex mixtures. Because of this, the Institute for Inland Water Management and Wastewater Treatment (RIZA), began developing a different approach for assessing effluents in a more thorough way.

Complex mixtures have become a focus for many researchers and policy-makers during the last few years. One reason for this is the astonishing number of substances which have been released into the environment. Well over 100 000 substances are being produced, used and emitted (Tonkes *et al.* 1995, EINECS). Therefore our land, air, surface waters, sediments and waste waters (or effluents) can now be considered as complex mixtures of both natural and anthropogenic substances, with many different properties. A problem we all face is limited knowledge about these substances. Many cannot be analysed, so there are often 'unknowns' present in a sample. Also, the properties of many substances that can be measured remain unknown. Are they toxic or bioaccumulative? Are they persistent or could they affect the genetic properties of organisms (including humans)? In most cases there is currently no answer. Another limitation is ignorance about the effect of combinations, or mixtures, of different substances. We rarely know the biological effect of a specific combination of substances. Attempts have been made to reduce this ignorance (e.g. Kraak 1992), but these are mostly focused on a small number of substances.

Forecasting the Environmental Fate and Effects of Chemicals. Edited by Philip S. Rainbow, Steve P. Hopkin and Mark Crane.
© 2001 John Wiley & Sons Ltd

Because of these limitations, it is not advisable to assess the potential environmental hazard of complex mixtures with a simple chemical or substance-specific approach. Additional approaches are necessary. One way to address the problem is a more effects-oriented approach, in which bioassays are used to assess effluents. Experience with such an approach for complex mixtures has already been established in different countries around the globe. The United States developed and implemented a whole effluent toxicity (WET) approach (USEPA 1991). Other examples of this approach can be found in Germany (Diehl and Hagendorf 1998) and the United Kingdom (Environment Agency 1996). Both of these European countries make use of acute toxicity testing to assess effluents. Germany implemented these tests more than 10 years ago. The UK is currently introducing an approach that is very similar to that of the US Environmental Protection Agency.

To understand the additional value, if any, of an effects-oriented approach, we must ask the question: 'Can a substance-specific chemical approach forecast the toxicity of effluents?' If it can, then the use of bioassays is redundant. This chapter describes work in the Netherlands designed to answer this question.

7.2   WHOLE EFFLUENT ASSESSMENT (WEA) IN THE NETHERLANDS

In the Netherlands water quality regulators currently perform substance-specific assessments for effluents. During the 1970s and the 1980s it became clear that this approach had many limitations. Unknown peaks were found in chemical analyses, but information on the characteristics of these 'unknowns' was not available. The number of substances found in effluents also increased, so that even if the properties of all substances present had been available, the costs for the analysis would have been prohibitive. Because of this it was found impossible to perform a complete environmental hazard assessment of effluent samples. RIZA therefore began the development of a different approach. The new effects-oriented approach, whole effluent assessment (WEA, formerly known as WEER or 'whole effluent environmental risk'), began in the early 1990s and makes possible a more complete assessment of effluents (Beckers-Maessen 1994, Tonkes and Botterweg 1994). Figure 7.1 shows the WEA approach, which comprises several different components: acute and chronic toxicity, genotoxicity, bioaccumulation, persistence and oxygen demand.

Because of extensive experience with acute toxicity testing, this component has been studied most thoroughly during the period 1994–1998 (De Graaf et al. 1996). Studies are currently ongoing for the other components and will result in possible implementation of WEA in the Netherlands around the year 2004.

7.2.1   ACUTE BIOASSAY APPROACH

Experience with acute toxicity testing in the Netherlands continues to grow. Studies with 76 large volume effluents originating from regional water authorities

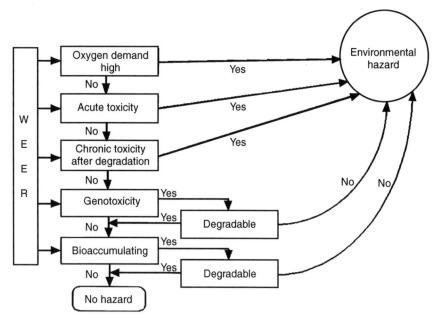

Figure 7.1   Whole effluent assessment scheme in the Netherlands.

have been performed. These effluents originate from industry and from municipal wastewater treatment plants. The industrial effluents that have been sampled are very diverse and include organic chemical and pharmaceuticals production, tank cleaning, metal processing and pesticide formulation. Biological testing has been performed with four different test organisms representing four trophic levels, i.e. bacteria, algae, crustaceans and fish. Organisms used were both freshwater and marine, and tests were performed according to international standards (Table 7.1). All effluents were chemically analysed and use was made of existing chemical data from consents and discharger reports.

7.2.2   COMPARISON OF ACUTE BIOASSAY AND CHEMICAL-SPECIFIC RESULTS

The results of the bioassays and the chemical analyses were compared with each other to answer the question: 'Can a substance-specific chemical approach forecast the toxicity of effluents?' The bioassay results represent the actual or **observed** toxicity of the complex effluent mixture. The chemical analyses should provide an expected or **predicted** toxicity. This predicted toxicity can be estimated by using toxicity data for specific chemicals from literature or databases (e.g. Aquire, developed by the US–EPA). A comparison such as this should provide an answer to our question and also clarify whether there is any additional value from using acute bioassays on effluents when compared with a substance-specific approach.

Toxic units (*TUs*) are calculated to make a quantitative comparison possible (Sprague and Ramsay 1965). First a *TU* is calculated for each substance found in an effluent based on chemical analysis and known toxicological data (equation 7.1):

$$TU_{substance} = \text{substance concentration}/E(L)C_{50substance} \qquad (7.1)$$

Then the assumption is made that all calculated *TUs* can be added, based on additive toxicity (no synergism or antagonism), to produce a total *TU* for all analysed substances for which literature or database data are available (equation 7.2).

$$TU_{substances} = TU_{substance1} + TU_{substance2} + TU_{substance3} \cdots + TU_{substance(n)} \qquad (7.2)$$

Then a *TU* is calculated for each bioassay result and each type of test organism (equation 7.3)

$$TU_{toxtest} = 100/E(L)C_{50} \text{(in volume percentage effluent, i.e. vol. \%)} \qquad (7.3)$$

Box 7.1 shows an example of how the separate *TUs* are produced according to this method.

Box 7.1   Simplified calculation of *TUs* for chemical analysis and toxicity test results with a crustacean (Tonkes and Baltus 1997)

Literature data for Daphnia magna EC_{50} *values*

$$Cu: EC_{50} = 10\,\mu g\,l^{-1};$$
$$As: EC_{50} = 1500\,\mu g\,l^{-1};$$

Concentrations of chemicals in effluent

$$Cu: 5\,\mu g\,l^{-1}: TU = 5/10 = 0.500$$
$$As: 5\,\mu g\,l^{-1}: TU = 5/1500 = 0.003$$

TU substances

$$TU_{Cu} + TU_{As} = 0.503$$

Toxicity test result

$$EC_{50} = 12 \text{ vol.\%}$$

TU_{toxtest}

$$100/EC_{50} = 100/12 = 8.33$$

Percentage toxicity explained by individual substances $= (0.503/8.33)*100\%$
$$= 6.04\%$$

Table 7.1 Test protocols for acute bioassays

Test organism	Freshwater	Saltwater
Bacteria Freshwater and marine: *Vibrio fischeri*	NVN 6516 ISO 11348	NVN 6516 ISO 11348
Algae Freshwater: *Raphidocelis subcapitata* Marine: *Phaeodactylum tricornutum/Skeletonema costatum*	OECD 201 ISO 8692 (and ISO 14442)	ISO 10253 (and ISO 14442)
Crustaceans Freshwater: water flea *Daphnia magna* Marine: copepod *Acartia tonsa*	OECD 202 ISO 6341	ISO 14669
Fish Freshwater: zebra fish *Danio rerio* Marine: guppy *Poecilia reticulata*	OECD 203 ISO 7346	OECD 203 ISO 7346

Comparison of chemical-specific and bioassay results is a simple step. The $TU_{substance}$ represents the **predicted** toxicity. The $TU_{bioassay}$ represents the **observed** toxicity. The predicted toxicity is divided by the observed toxicity and multiplied by 100%. The outcome is the explained percentage of the toxicity found in the effluent sample. Initial results (De Graaf *et al.* 1996, 2000, Tonkes and Baltus 1997) in general show that there is a large gap between predicted and observed toxicity results. In many cases ($> 50\%$ of the effluent samples) only 20% or less of the toxicity test results can be explained by the known substances for which toxicity data are available. Tables 7.2–7.4 illustrate some of these initial results.

Table 7.2 Example of toxicity test results with a chemical industry effluent (results in vol.%)

Bacteria (*Vibrio fischeri*)	$EC_{20} => 45$
Algae (*Phaeodactylum tricornutum*)	$EC_{50} = (A^a) = 55$
	$EC_{50} = (\mu^b) = > 80$
Crustaceans (*Acartia tonsa*)	$EC_{50} = 57$
Fish (*Poecilia reticulata*)	$LC_{50} = > 100$

[a] Surface area under the growth curve, i.e. biomass.
[b] Growth rate.

Table 7.3 Chemical analysis data and TUs for several substances

Substance	Concentration	Bacteria		Crustaceans	
		EC_{50}	TU	EC_{50}	TU
As	$8 \mu g\,l^{-1}$	4 900	0.002	1 500	0.005
Pb	$23 \mu g\,l^{-1}$	—		450	0.051
o-xylene	$5 \mu g\,l^{-1}$	9 200	0.001	3 200	0.002
Toluene	$3 \mu g\,l^{-1}$	—		7 000	0.0004
Chloroform	$12 \mu g\,l^{-1}$	52 000	0.0002	78 900	0.0002

Table 7.4 Example of comparison for a chemical industry effluent

Test organism	$TU_{substances}$	$TU_{bioassay}$	% explained	% data known[a]
Bacteria	1.5	—	[b]	16
Algae	0.05	1.8	2.5	14
Crustaceans	2.3	1.8	>100	44
Fish	0.26	—	[b]	7

[a] Percentage of the total amount of known substances in an effluent for which ecotoxicological data are available.
[b] Toxicity expected, but not found.

7.3 CAN A SUBSTANCE-SPECIFIC CHEMICAL APPROACH FORECAST THE TOXICITY OF EFFLUENTS?

The previous section showed that a large percentage of the toxic effects in the tested effluent samples cannot be explained simply by separate substance toxicity. Therefore, an accurate forecast of the effects of toxic chemicals based only on chemical analysis is not possible. There are several reasons why this conclusion may not be valid. The large discrepancy between observed and predicted toxicity may be caused by unknown substances that have not been analysed, or by substances for which no toxicity data are available. If those unknowns can be analysed and data become available, more accurate forecasts may be possible. However, it is doubtful whether this is realistic when one considers the enormous number of substances potentially present in an effluent. Another possible explanation for the large discrepancy between observed and predicted is that the combination of known substances has led to the observed toxicity through synergism. Tests of this hypothesis are theoretically possible, but the number of substances in complex effluent mixtures would make such tests very difficult and perhaps impossible in practice. A final reason for the difference in observed:predicted toxicity may be that calculation of TUs influenced the conclusion. The TU method has several limitations including:

- lack of known data for substances which can be analysed
- unreliable literature data
- matrix effects (bioavailability)
- influence of unknowns
- assumption of additivity

However, at least the first three limitations are also valid for the assessment of separate substances.

Because WEA will probably be implemented in the Netherlands it is important to establish whether the other components of the scheme (chronic toxicity, genotoxicity, bioaccumulation, persistence) are also of additional value in comparison to a substance-specific approach. Therefore the next challenge is to make a quantitative comparison possible for these other components. Whether this is practically possible, remains to be seen.

REFERENCES

Beckers-Maessen CMH (1994) *Toxiciteitstesten in de Wvo-vergunningverlening.* RIZA-werkdocument 94.071X (in Dutch) 31 pp.

De Graaf PJF, Graansma J, Tonkes M, ten Kate EV and Beckers CMH (1996) *Acute Toxicity Tests. An Addition to the Specific Substances Approach? Study of Industrial Effluents in North-Netherlands and South-Holland.* FWVO-Report 96.03, 66 pp. incl. appendix (in Dutch).

De Graaf PJF, Graansma J, Tonkes M, Maas H and ten Kate EV (2000) *Acute Toxicity Tests Performed on Complex Effluents. Results with Effluent Samples from Regional Directorates.* FWVO-report, in press.

Diehl K and Hagendorf U (1998) *Datensammlung Bioteste. Erhebungen, Bewertung, Empfehlungen.* Institut für Wasser-, Boden- und Lufthygiene des Umweltbundesamtes. Texte 9-98. ISSN 0722-186X (in German).

Environment Agency (1996) *The Application of Toxicity-based Criteria for the Regulatory control of Wastewater Discharges. A Consultation Document issued by the Environment Agency on the Proposed Introduction of Toxicity-based Conditions to Discharge Licences.* Environment Agency, Worthing, UK, 50 pp.

ISO 6341 (1989) *Water Quality— Determination of the Inhibition of the Mobility of* Daphnia magna *Straus* (Cladocera, Crustacea). International Organization for Standardization.

ISO 8692 (1989) *Water Quality— Freshwater Algal Growth Inhibition Test with* Scenedesmus subspicatus *and* Selenastrum capricornutum. International Organization for Standardization.

Kraak MHS (1992) *Ecotoxicity of metals to the freshwater mussel* Dreissena polymorpha. PhD thesis. University of Amsterdam, Amsterdam, the Netherlands.

NVN 6516 (1993) *Water— Bepaling van de acute toxiciteit met behalf van* Photobacterium phosphoreum. Nederlands Normalisatie Instituut, Delft, augustus 1993 (Dutch protocol for bacteria test).

OECD 202 part II (1993) *Draft OECD Test Guideline 202 Part II,* Daphnia magna *Reproduction Test to be Used in the Final Ring Test.* Organization for Economic Co-operation and Development, Paris, France.

OECD 203 (1992) *Fish, Acute Toxicity Test. OECD Guideline for Testing of Chemicals.* Organization for Economic Co-operation and Development, Paris, France, 203 adopted 17.07.92.

OECD 210 (1992) *Fish, Early-Life Stage Toxicity Test. OECD Guideline for Testing of Chemicals.* Organization for Economic Co-operation and Development, Paris, France, 210 adopted 17.07.92.

Sprague JB and Ramsay BA (1965) Lethal levels of mixed copper-zinc solutions for juvenile salmon. *Journal Fisheries Research Board of Canada,* **22**, 425–432.

Tonkes M and Baltus CAM (1997) *Praktijkonderzoek aan complexe effluenten met de Totaal Effluent Milieubezwaarlijkheid (TEM)— methodiek. Resultaten van 10 complexe effluenten.* RIZA-rapport 97.033. SPA-rapport 97.003, 78 pp. excl. Appendix (in Dutch with English summary).

Tonkes M and Botterweg J (1994) *Totaal Effluent Milieubezwaarlijkheid; Beoordelingsmethodiek milieubezwaarlijkheid van afvalwater. Literatuur—en gegevensevaluatie.* RIZA, notanr. 94.020, 157 pp. (in Dutch with English summary).

Tonkes M, van de Guchte C, Botterweg J, de Zwart D and Hof M (1995) *Monitoring Water Quality in the Future. Volume 4: Monitoring Strategies for Complex Mixtures.* Report No. 4 in a series of 6. 104 pp.

USEPA (1991) *Technical Support Document for Water Quality-Based Toxics Control.* United States Environmental Protection Agency, Office of Water, Washington, DC. EPA/505/2-90-001, 45 pp.

Forecasting the Environmental Effects of Zinc, the Metal of Benign Neglect in Soil Ecotoxicology

STEVE P. HOPKIN[1] AND DAVID J. SPURGEON[2]

[1]*Division of Zoology, School of Animal and Microbial Sciences, University of Reading, Reading, UK*
[2]*Institute of Terrestrial Ecology, Monks Wood, Abbots Ripton, Huntingdon, Cambs, UK*

8.1 INTRODUCTION

It was over 20 years ago that Stubbs (1977) published his short paper which drew attention to the then neglected metal cadmium and its importance as an environmental pollutant. Since then, several international conferences and many thousands of papers have been published on the biochemistry, toxicology and ecotoxicology of cadmium (see e.g. Stoeppler 1991). It is our view that soil ecotoxicologists have become preoccupied with this element. Apart from a few recent exceptions (e.g. Donker *et al.* 1996, 1998, Drobne and Strus 1996, Hopkin and Hames 1994, Smit and van Gestel 1996, 1997, 1998, Smit *et al.* 1997, 1998, Spurgeon and Hopkin 1996a,b,c; 1999, van Gestel and Hensbergen 1997), most workers have neglected to consider the contribution to overall toxicity made by zinc which is almost always associated with cadmium in soils, but at much higher concentrations.

Zinc is an essential element for all organisms and the general view is that it is 'non-toxic' (Ohnesorge and Wilhelm 1991, Vallee and Falchuk 1993). For example, emissions of cadmium from incinerators are closely monitored whereas zinc is hardly considered at all (Carnes *et al.* 1992, Hutton *et al.* 1988). Most studies have emphasized the importance of zinc deficiency in terrestrial ecosystems rather than excess (e.g. Boardman and McGuire 1990b). Consequently, relatively few experiments have been conducted on the toxicity of zinc to terrestrial invertebrates. For example, van Straalen (1993) was unable to propose 'protection levels' for zinc in soil due to the lack of published experimental data on which to base his models.

Cadmium is present as a by-product of zinc smelting and occurs at about 0.3% of zinc production. Aerial emissions and subsequent deposition of cadmium during

Forecasting the Environmental Fate and Effects of Chemicals. Edited by Philip S. Rainbow, Steve P. Hopkin and Mark Crane.
© 2001 John Wiley & Sons Ltd

smelting are typically between one-fiftieth and one-hundredth of those of zinc (Ewers 1991, Harrison *et al.* 1993, Martin and Bullock 1994, Martin *et al.* 1982), except at a handful of rare sites where secondary smelting of cadmium occurs (e.g. the sediments studied by Klerks and Levinton 1993). Few studies have **simultaneously** compared the toxicities of cadmium and zinc to terrestrial animals. Cadmium has invariably been found to be more toxic than zinc when effects on growth, reproduction and survival in terrestrial invertebrates are considered (Bengtsson and Tranvik 1989, Hopkin 1989, Hopkin and Hames 1994, Jørgensen *et al.* 1991).

Ecotoxicologists are concerned with studying the effects of contaminants on ecosystems. Two approaches are possible. First, the scientist can survey contaminated field sites to observe pollution damage. Second, laboratory experiments can be attempted to predict field effects (for a comprehensive review of such tests, see Løkke and van Gestel 1998). However, there are problems with these approaches. How does one determine which chemical among a 'cocktail' of pollutants is most likely to be responsible for ecosystem damage? Also, are toxicity parameters determined in laboratory experiments ($EC_{50}s$, $LC_{50}s$ etc.) comparable with effects observed in real habitats?

In this chapter, we show how standard laboratory tests can be used for determining **relative** toxicities of metals with consequences for the forecasting of potential toxic effects. The toxicity ratios determined allow conclusions to be drawn as to which metal within a 'cocktail' of contaminants is likely to be responsible for reduced performance of invertebrates in polluted soils.

8.2 AVONMOUTH

The Avonmouth primary Zn–Cd–Pb smelting works in SW England is one of the largest industrial plants of its type in the world. The effects of emissions of metals on surrounding ecosystems have been studied extensively (for reviews see Hopkin 1989, Martin and Bullock 1994, Walker *et al.* 1996). Soils in the surrounding region are heavily contaminated with cadmium, copper, lead and zinc. Soil invertebrates are scarce in the vicinity of the factory and a thick layer of undecomposed leaf litter has built up on the soil surface due mainly to the lack of earthworms (Spurgeon and Hopkin 1996a). However, because the metals occur together in a 'cocktail', it is difficult to decide which element is primarily responsible for the absence of detritivorous animals.

One approach to solve this problem is to compare the relative toxicities of the four metals in laboratory experiments and to relate these values to concentrations of cadmium, copper, lead and zinc in soils near to the factory. In this context, published data from experiments conducted in our laboratory in Reading have been reanalysed and a term, the **relative toxicity factor** is introduced. For further details of how the tests are conducted, see Hopkin (1997) and Walker *et al.* (1996) and the original papers cited at the foot of Table 8.1.

Table 8.1 Calculation of T_F for four species of soil invertebrate with reference to soil metal concentrations at Avonmouth and toxicity parameters ($\mu g\, g^{-1}$ dry weight in soil or food)

	Cd	Cu	Pb	Zn
Soil 3 km from smelter $\mu g\, g^{-1}$	10	50	500	1000
C_{Cd}	1	5	50	100
Eisenia fetida[a] (soil)				
14-day EC$_{50}$ cocoon prodn	295	716	1629	357
T_{Cd}	1	2.4	5.5	1.2
T_F	1	2.1	9.1	83
Helix aspera[b] (food)				
42-day EC$_{50}$ reprodn	183	1050	8300	5970
T_{Cd}	1	5.7	45	33
T_F	1	0.9	1.1	3.0
Folsomia candida[c] (soil)				
42-day EC$_{50}$ juvenile prodn	540	640	1570	590
T_{Cd}	1	1.2	2.9	1.1
T_F	1	4.2	17	91
Porcellio scaber[d] (food)				
360-day LC$_{50}$ juvenile survival	40	60	1200	700
T_{Cd}	1	1.5	30	17
T_F	1	3.3	1.7	5.6

[a] 20 °C: Spurgeon and Hopkin (1995).
[b] 15 °C: Laskowski and Hopkin (1996).
[c] 15 °C: Sandifer and Hopkin (1997).
[d] 20 °C: Hopkin and Hames (1994).

Box 8.1 Procedure for determining the relative toxicity factor (T_F) in soil invertebrates for Cd, Cu, Pb and Zn in contaminated soils at Avonmouth, SW England

1. Determine concentration of each metal in fields soils (Cd, Cu, Pb, Zn).
2. Calculate concentration of each metal relative to cadmium (C_{Cd}).

$$C_{Cd} = \frac{\text{concentration of metal in soil } (\mu g\, g^{-1} \text{ dry weight})}{\text{concentration of cadmium in soil } (\mu g\, g^{-1} \text{ dry weight})}$$

3. Determine toxic concentration of each metal in food or soil affecting a specific parameter for the species in question using laboratory test (e.g. LC$_{50}$, EC$_{50\text{reproduction}}$).
4. Calculate toxicity of each metal relative to cadmium (T_{Cd}).

$$T_{Cd} = \frac{\text{value of toxicity parameter for metal } \mu g\, g^{-1} \text{ dry weight}}{\text{value of toxicity parameter for cadmium } \mu g\, g^{-1} \text{ dry weight}}$$

5. Calculate relative toxicity factor for each metal relative to its concentration in field soils:

$$\text{Relative toxicity factor } T_F = \frac{C_{Cd}}{T_{Cd}}$$

6. The metal with the highest T_F value is the one most likely to be causing toxic effects in the field.

8.3 RELATIVE TOXICITY FACTOR

The method for determining the relative toxicity factor (T_F) is shown in Box 8.1. This approach has been applied to several species of soil invertebrate of which four examples are shown in Table 8.1. These are the OECD 'standard earthworm' *Eisenia fetida*, the common garden snail, *Helix aspersa*, the 'standard springtail' *Folsomia candida* (Insecta: Collembola) and the terrestrial isopod, *Porcellio scaber*. A unique feature of these experiments is that the toxicities of cadmium, copper, lead and zinc were determined simultaneously for all test organisms.

Results from these four species confirm that the highest values for T_F are always obtained for Zn in relation to Avonmouth soils. Although some experiments have demonstrated interactions between the metals when they are presented to the test animals in mixtures, reduced or enhanced toxicities due to synergism or antagonism are relatively insignificant. The small reduction or enhancement of the toxicity of individual metals which occurs does not invalidate the conclusion that to protect primary consumers at Avonmouth, emission controls should be targeted at zinc. Cadmium and lead are regarded with much greater concern by regulators, but it is clear that the toxic effects of zinc need to be taken much more seriously. The importance of zinc has also been highlighted by Boardman and McGuire (1990a), Laskowski *et al.* (1994) and van Assche and Clijsters (1990).

8.4 CONCLUSIONS

1. Standard laboratory tests can be used to provide useful information on the relative toxicities of chemicals to soil invertebrates, enabling the identification of the most damaging chemical within a 'cocktail' of pollutants present in a soil. Such data can then be used to forecast the potential effects of the constituent pollutant chemicals on soil invertebrates.
2. At Avonmouth, it appears that zinc is the metal which is causing most damage to detrivorous soil invertebrates, especially earthworms.
3. Zinc must be included in any legislation concerned with soil protection as its importance as an environmental pollutant has been severely underestimated.

ACKNOWLEDGEMENTS

This work was supported by grants from the Leverhulme Trust and the Natural Environment Research Council.

REFERENCES

Bengtsson G and Tranvik L (1989) Critical metal concentrations for forest soil invertebrates. *Water, Air and Soil Pollution*, **47**, 381–417.

Boardman R and McGuire DO (1990a) The role of zinc in forestry. I. Zinc in forest environments, ecosystems and tree nutrition. *Forestry Ecology and Management*, **37**, 167–205.

Boardman R and McGuire DO (1990b) The role of zinc in forestry. II. Zinc deficiency and forest management: effect on yield and silviculture of *Pinus radiata* plantations in South Australia. *Forestry Ecology and Management*, **37**, 207–218.

Carnes RA, Santoleri JJ and McHale SH (1992) Metals and incinerators: the latest regulatory phase. *Journal of Hazardous Materials*, **30**, 343–353.

Donker MH, Raedeckler MH and van Straalen NM (1996) The role of zinc regulation in the zinc tolerance mechanism of the terrestrial isopod *Porcellio scaber. Journal of Applied Ecology*, **33**, 955–964.

Donker MH, Abdel-Lateif HM, Khalil MA, Bayoumi BM and van Straalen NM (1998) Temperature, physiological time, and zinc toxicity in the isopod *Porcellio scaber. Environmental Toxicology and Chemistry*, **17**, 1558–1563.

Drobne D and Strus J (1996) Molt frequency of the isopod *Porcellio scaber* as a measure of zinc-contaminated food. *Environmental Toxicology and Chemistry*, **15**, 126–130.

Ewers U (1991) Standards, guidelines and legislative regulations concerning metals and their compounds. In *Metals and their Compounds in the Environment*, Merian E (ed.), VCH, Weinheim, pp. 687–711.

Harrison SJ, Vale JA and Watts CD (1993) The estimation of aerial inputs of metals to estuarine waters from point source pattern data using an isoplething technique: Severn Estuary, U.K. *Atmospheric Environment*, **27A**, 2337–2349.

Hopkin SP (1989) *Ecophysiology of Metals in Terrestrial Invertebrates*. Elsevier Applied Science, 366 pp.

Hopkin SP (1997) *Biology of the Springtails (Insecta: Collembola)*. Oxford University Press, 330 pp.

Hopkin SP and Hames CAC (1994) Zinc, among a 'cocktail' of metal pollutants, is responsible for the absence of the terrestrial isopod *Porcellio scaber* from the vicinity of a primary smelting works. *Ecotoxicology*, **2**, 68–78.

Hutton M, Wadge A and Milligan PJ (1988) Environmental levels of cadmium and lead in the vicinity of a major refuse incinerator. *Atmospheric Environment*, **22**, 411–416.

Jørgensen SE, Nielsen SN and Jørgensen LA (1991) *Handbook of Ecological Parameters and Ecotoxicology*. Elsevier Science, Amsterdam.

Klerks PL and Levinton JS (1993) Evolution of resistance and changes in community composition in metal-polluted environments: a case study on Foundry Cove. In *Ecotoxicology of Metals in Invertebrates*, Dallinger R and Rainbow PS (eds), Lewis, Chelsea, USA, pp. 223–240.

Laskowski R and Hopkin SP (1996) Effects of Zn, Cu, Pb and Cd on fitness in snails (*Helix aspersa*). *Ecotoxicology and Environmental Safety*, **34**, 59–69.

Laskowski R, Maryanski M and Niklinska M (1994) Effect of heavy metals and mineral nutrients on forest litter respiration rate. *Environmental Pollution*, **84**, 97–102.

Løkke H and van Gestel CAM (eds) (1998) *Handbook of Soil Invertebrate Toxicity Tests*. Wiley, Chichester.

Martin MH and Bullock RJ (1994) The impact and fate of heavy metals in an oak woodland ecosystem. In *Toxic Metals in Soil-Plant Systems*, Ross SM (ed.), Wiley, Chichester, UK, pp. 327–365.

Martin MH, Duncan EM and Coughtrey PJ (1982) The distribution of heavy metals in a contaminated woodland ecosystem. *Environmental Pollution*, **3B**, 147–157.

Ohnesorge FK and Wilhelm M (1991) Zinc. In *Metals and their Compounds in the Environment*, Merian E (ed.), VCH, Weinheim, pp. 1309–1342.

Sandifer RD and Hopkin SP (1997) Effects of temperature on the relative toxicities of cadmium, copper, lead and zinc to *Folsomia candida* Willem, 1902 (Collembola) in a standard laboratory test system. *Ecotoxicology and Environmental Safety*, **37**, 125–130.

Smit CE and van Gestel CAM (1996) Comparison of the toxicity of zinc for *Folsomia candida* in artificially contaminated and polluted field soils. *Applied Soil Ecology*, **3**, 127–136.

Smit CE and van Gestel CAM (1997) Influence of temperature on the regulation and toxicity of zinc in *Folsomia candida* (Collembola). *Ecotoxicology and Environmental Safety*, **37**, 213–222.

Smit CE and van Gestel CAM (1998) Effects of soil type, prepercolation, and ageing on bioaccumulation and toxicity of zinc for the springtail *Folsomia candida. Environmental Toxicology and Chemistry*, **17**, 1132–1141.

Smit CE, van Beelen P and van Gestel CAM (1997) Development of zinc bioavailability and toxicity for the springtail *Folsomia candida* in an experimentally contaminated field plot. *Environmental Pollution*, **98**, 73–80.

Smit CE, van Overbeck I and van Gestel CAM (1998) The influence of food supply on the toxicity of zinc for *Folsomia candida* (Collembola). *Pedobiologia*, **42**, 154–164.

Spurgeon DJ and Hopkin SP (1995) Extrapolation of the laboratory-based OECD earthworm toxicity test to metal-contaminated field sites. *Ecotoxicology*, **4**, 190–205.

Spurgeon DJ and Hopkin SP (1996a) The effects of metal contamination on earthworm populations around a smelting works: quantifying species effects. *Applied Soil Ecology*, **4**, 147–160.

Spurgeon DJ and Hopkin SP (1996b) Effects of variation of the organic matter content and pH of soils on the availability and toxicity of zinc to the earthworm *Eisenia fetida. Pedobiologia*, **40**, 80–96.

Spurgeon DJ and Hopkin SP (1996c) Effects of metal-contaminated soils on the growth, sexual development and early cocoon production of the earthworm *Eisenia fetida* with particular reference to zinc. *Ecotoxicology and Environmental Safety*, **35**, 86–95.

Spurgeon DJ and Hopkin SP (1999) Tolerance to zinc in populations of the earthworm *Lumbricus rubellus* from uncontaminated and metal-contaminated ecosystems. *Archives of Environmental Contamination and Toxicology*, **37**, 332–337.

Stoeppler M (1991) Cadmium. In *Metals and their Compounds in the Environment*, Merian E (ed.), VCH, Weinheim, pp. 803–851.

Stubbs RL (1977) Cadmium — the metal of benign neglect. In *Cd 77: Proceedings of the First International Cadmium Conference*. San Francisco Metal Bulletin, Drogher Press, Dorset, England, pp. 7–12.

Vallee BL and Falchuk KH (1993) The biochemical basis of zinc physiology. *Physiological Reviews*, **73**, 79–118.

van Assche F and Clijsters H (1990) A biological test system for the evaluation of the phytotoxicity of metal-contaminated soils. *Environmental Pollution*, **66**, 157–172.

van Gestel CAM and Hensbergen PJ (1997) Interaction of Cd and Zn toxicity for *Folsomia candida* Willem (Collembola: Isotomidae) in relation to bioavailability in soil. *Environmental Toxicology and Chemistry*, **16**, 1177–1186.

van Straalen NM (1993) Soil and sediment quality criteria derived from invertebrate toxicity data. In *Ecotoxicology of Metals in Invertebrates*, Dallinger R and Rainbow PS (eds), Lewis, Chelsea, USA, pp. 427–441.

Walker CH, Hopkin SP, Sibly RM and Peakall DB (1996) *Principles of Ecotoxicology*. Taylor and Francis, 321 pp.

9

Forecasting in an Uncertain World: Managing Risks or Risky Management?

DAVID SANTILLO, PAUL JOHNSTON AND RUTH STRINGER

Greenpeace Research Laboratories, University of Exeter, Exeter, UK

9.1 INTRODUCTION: PREDICTION AND FORECASTING

A capacity to identify environmental threats in advance, combined with a strategy to prevent serious or irreversible degradation, are essential to ensure effective environmental protection and sustainable use of ecosystem services. Such a capacity requires an approach which permits, at some level, the identification and description of the possible outcomes of any particular activity, their likelihood and their potential magnitude or severity. In other words, decisions regarding future conditions must be made on the basis of derivation and deduction from current or historical conditions and trends.

Attempts to provide answers to questions regarding the future fall into the realms of prediction. Predictions, whether qualitative or quantitative, therefore have a number of applications:

- to provide information on which to base decisions relating to the future
- to enable adaptations to foreseen changes over which we have no control

and, more specifically to environmental management,

- to ensure that current exploitation of ecosystem services is sustainable, and
- to avoid environmental problems, or the potential for problems, before they occur.

Forecasting, though often used synonymously with prediction, is, more accurately, the process of arriving at predictions which are unconditional, specific and absolute (Rescher 1998). Forecasts should also strictly be verifiable with respect to their accuracy at some point in the future.

The mathematical theory of prediction, and of forecasting specifically, is complex and reflects a continuum of hypotheses ranging from the predeterministic to the chaotic. The mathematician Pierre Simon de Laplace hypothesized that if one could describe in perfect detail the condition, dynamics and underlying mechanisms of a system, then it should be possible to calculate future conditions with complete certainty. In practice, of course, prediction and forecasting are less certain disciplines. It is, after all, impossible to know precisely

Forecasting the Environmental Fate and Effects of Chemicals. Edited by Philip S. Rainbow, Steve P. Hopkin and Mark Crane.
© 2001 John Wiley & Sons Ltd

what the future holds until such time as it is no longer the future. At that time it may be possible to review the accuracy of the predictions, and modify future predictions accordingly, but predictive certainty will never be achieved.

Despite the fact that prediction is an inherently uncertain pursuit, substantial resources are invested in predictive practices, and still more depend on their accuracy. Sherden (1998) lists seven established predictive professions, including meteorology, economic forecasting, investment management, demography and organizational planning. In most cases, there is a heavy reliance on quantitative models, generated and validated on the basis of historical data sets. Other less professional predictive activities, including astrology and fortune telling, are perhaps less mechanistic and rely to a significant degree on the predictions being unverifiable or, at least, somewhat open to interpretation.

When preparing forecasts on issues of environmental protection, for the purpose of ensuring that policy measures are protective, ambiguity of predicted outcome may be wholly unacceptable as consequences may potentially be widespread and severe. While it is clearly reasonable to strive for forecasts which are as accurate as possible, recognition of uncertainties in the information and model on which the forecast is based, and of the possible consequences of the forecast being inaccurate, become equally as important. Clearly, as the severity of the consequences increases, not only does the need for improvements in predictive power increase, but also the need to minimize the possibility that a forecast-based decision may later prove incorrect. Indeed, the latter consideration may be most important when the facts are very uncertain and the decision stakes high.

9.2 ECOSYSTEM COMPLEXITY AND 'THE REGULATOR'S DILEMMA'

With respect to forecasting of environmental pathways and impacts, it must be recognized that ecosystems are complex entities which can neither be defined explicitly nor described fully, other than through imposition of artificial boundaries and the consequent generation of unquantified externalities. Berg and Scheringer (1994) derive the term 'overcomplexity' to describe the manner in which the spatial relationships and temporal evolution of ecosystems are both unpredictable and impossible to examine to a degree sufficient to reveal their detail and 'derivative properties' (Power and McCarty 1997). Furthermore, meaningful and representative reference conditions, including variability in time and space, against which ecosystem stresses and damage can be gauged are likely to be impossible to discern, a characteristic which Berg and Scheringer (1994) term 'normative indeterminacy'.

It is this background against which regulatory decisions are required in order to avoid systematic environmental degradation. With regard to the limitation of damage to ecosystems and the maintenance of their viability through protection or sustainable exploitation, it has long been recognised that 'prevention is better than cure' (Bodansky 1991). However, identifying the optimal point at which to take preventative action is undoubtedly a difficult task. In common regulatory practice there appears to be little will to take truly precautionary preventative action because of the possibility that the severity of potential problems may later prove to have been overestimated, leading to measures which may, therefore, have been unnecessarily overprotective. Nevertheless, to do otherwise, in the hope that environmental degradation will be identified and related to specific factors which may then be addressed before such damage becomes severe or even irreversible, is to rely on a number of assumptions which may be fundamentally flawed. Conferring the benefit of any doubt over impacts on to the activity or chemical in question may be viewed, in general terms, as a less responsible approach to environmental protection.

Clearly, the deferral of a decision until such time as a potentially impacted system can be fully described, and the consequences of a particular stressor reliably quantified or predicted, is a decision in itself. Furthermore, it must be recognized that such analytical certainty will never be achieved and that serious or irreversible damage may result while greater certainty concerning cause and effect relationships is sought. Indeed, while clearly invaluable for improved understanding of ecosystem function, research frequently demonstrates that the definition of ecosystem processes is a more difficult task than initially assumed. Environmental regulators are, therefore, frequently presented with the need to take action to prevent, or avoid the potential for, damage to the environment or human health in the face of considerable uncertainty, an unquantifiable degree of ignorance and inherent indeterminacies, a situation Bodansky (1991) terms the 'regulator's dilemma'.

Numerous approaches have emerged in attempts to resolve, or avoid, this dilemma. This chapter explores in more detail two contrasting approaches:

1. Risk-based approaches, which use scientific information in attempts to arrive at system descriptions and predictions of effects which are as numerically accurate as possible, assuming that reliability of predictions can be increased progressively through reduction and elimination of uncertainties; and
2. A precautionary approach, which uses scientific information in order to guide the development of preventative measures, recognizing inherent indeterminacies and limitations to knowledge and ensuring that such measures err on the side of environmental protection.

Both approaches rely on the provision of high-quality scientific information, but differ in the underlying philosophies governing its interpretation.

9.3 LIMITATIONS OF RISK-BASED APPROACHES

Approaches based on the assessment and management of risk have perhaps received the greatest attention in recent years, relying essentially on the application of techniques developed in engineering sciences to the forecasting of trends and impacts in more complex and poorly defined natural systems. Risk-based approaches extend from the view that environmental risks can first be accurately quantified and subsequently managed at levels which may be considered sustainable and 'acceptable', either in absolute terms or relative to the benefits derived. In broad terms, risks are determined from a combination of the intrinsic hazards presented by a chemical or activity and measurements or estimates of exposure to that agent.

In recognition of their mechanistic derivation, it may be argued that risk assessment and management have an underlying philosophy which is essentially Laplacian in nature. Such approaches assume that it is possible to know enough about the hazards of, and exposure to, a particular chemical or activity to enable calculation of the risks in a reliable manner. Risk assessment is viewed as a suite of techniques which can have universal application to the evaluation of diverse threats to ecosystems and human health. Furthermore, the central importance of scientific determinations or estimations of hazards and of exposure to these hazards is seen to confer to risk assessment a high degree of objectivity and rigour (Burger and Gochfeld 1997).

Scientific research undoubtedly has the ability to improve our understanding of ecosystem function and the interrelationships of component media, organisms and pathways and may, consequently, allow the reduction of a number of recognized uncertainties. Nevertheless, all scientific determinations are bound by the largely arbitrary constraints of experimental design and the need to control, as far as possible, all externalities. Further analysis of a particular system can never address those properties about which we remain ignorant, other than if such properties are identified by chance and are then amenable to reduction. Moreover, the inherent existence of processes, networks and chains of causality within natural systems which are essentially open mean that substantial irreducible uncertainties or indeterminacies (Wynne 1992, Dovers and Handmer 1995) will always remain as barriers to comprehensive description and prediction of ecosystem function.

The implicit and operational limitations, uncertainties and indeterminacies of the description of ecosystem pathways and interactions can result, in turn, in the transmission of substantial uncertainty and inaccuracy to the analysis of associated risk. Despite the fact that failure to recognize and account for such unknowns can have severe consequences, uncertainties and indeterminacies arising from ecosystem complexity are rarely made explicit. Simple numerical expressions of risk (e.g. the ratio of predicted environmental concentration to the predicted no-effect concentration, the PEC/PNEC ratio) remain in widespread use, although they are clearly very limited in the information they convey and, possibly, grossly misleading if separated from the specific assumptions on which they are based (Ohanian *et al.* 1997). Green and Crouch (1997) caution that the outcome of risk assessments will always depend on an unquantifiable degree on details which are inherently unknowable.

The PEC/PNEC risk assessment approach is central to current instruments for the regulation of chemicals within the European Union (as set out in the *Technical Guidance Documents* relating to Directive 93/67/EEC and Regulation 1488/94 for new and existing chemicals respectively). Essentially, forecasts of the fates and effects of a particular chemical are conducted on the basis of predictions of the concentrations to which individuals in a particular environmental compartment are exposed and the concentrations which may be expected to have effects, for what are nominally the 'most sensitive' species and end points. In reality, of course, the criteria are merely the most sensitive of those included within the assessment, under the particular conditions employed in the underlying studies. Moreover, a great deal of detailed information from the routine screening of chemicals or from primary research, including the assumptions upon which that information is contingent, may be lost during the distillation of data down to a simple ratio of theoretical constructs. The entire decision-making process then turns upon whether or not the PEC/PNEC ratio is >1 or <1. If <1, no further consideration or action may be required. The levels of uncertainties which pervade such estimates, and which could, in theory, result in the boundaries of confidence straddling this cut-off value, are almost invariably not made explicit.

Despite the simplistic nature of the final assessment, for the majority of chemicals even the basic data sets required to conduct an assessment are not yet available. Furthermore, for those chemicals which have been prioritized for risk assessment few have been completed, and measures to address even a single one have yet to be implemented. Moreover, while assessments which may well subsequently identify problems are ongoing, provisions are rarely made for interim precautionary action. These issues have been well reviewed by the European Environment Agency (EEA 1998).

As risk management is predicated upon reliable evaluation of risk, these measures will also be subject to the same limitations as outlined above. However, in contrast to risk assessment, the fact that risk management is pervaded by inherent subjectivity and value judgement is commonly recognized (Gerard and Petts 1998). As noted by Berg and Scheringer (1994), 'acting in an environmentally sustainable way is more than just applying ecological knowledge'. A risk deemed acceptable according to ectotoxicological assessment and defined environmental acceptability criteria may, nevertheless, be unacceptable in political or social terms (Bewers 1995). The size of the decision stakes and the potential consequences of being wrong are also essential considerations when arriving at policy decisions (Wynne 1992). Moreover, issues of inequity may also arise when the populations or environmental compartments deriving benefit from the chemical or activity in question are distinct from those which bear the consequences (EEA 1998). In addition, the question also remains as to who would bear responsibility for a decision based on an assessment of risk subsequently found to be incorrect (Funtowicz and Ravetz 1994).

Risk assessment and management may, therefore, not only be limited by poor data availability (a commonly identified problem) and uncertainties and errors in measurements, but also more fundamentally by the degree to which ecosystem processes may be understood and modelled for the purposes of forecasting. The existence of both indeterminacies (irreducible uncertainties) and ignorance of pathways and/or interactions, can lead to situations in which models which are complex and well furnished with data nevertheless fail to predict even within very broad margins of error (e.g. see Box 9.1).

Despite the limitations identified above, it is generally assumed that risk-based quantitative forecasting will ultimately provide the central basis for environmental legislation. Clearly there is an ongoing need for pure and applied knowledge to facilitate improved understanding of ecosystem function and contaminant behaviour (Ducrotoy and Elliot 1997), but such knowledge will always be incomplete and, in itself, can form only part of responsible policy and management systems. Scientific analyses and deductions undoubtedly serve policy-makers with valuable information, and this can be used to identify hazards and to prioritize and guide decisions of a more precautionary nature (Funtowicz and Ravetz 1994). Fundamentally, however, such assessments cannot replace the decision-making process itself (Wynne 1992, Power and McCarty 1997). Despite this, the calculation of simple PEC/PNEC ratios is commonly viewed as the fulcrum on which the identification of the need for further measures turns.

These issues are discussed in more detail elsewhere (Johnston *et al.* 1996, Santillo *et al.* 1998, 2000). Nevertheless, the possible consequences of such limitations to the scope of assessments may be well illustrated by an example drawn from a discipline other than chemical contamination, that of fisheries management (Box 9.1).

Box 9.1   Case study: The collapse of the Canadian cod stocks

Between 1960 and 1992 the spawning stock biomass of cod on the eastern seaboard of Canada fell from 1.6 Mt to only 22 000 t leading to the complete closure of the fishery in 1992 (Hutchings and Myers 1994, Hutchings 1996). Since then, recovery of stocks has been limited by this exceptionally low spawning stock biomass. This collapse occurred in spite of intensive management of the stocks, based on the application of sophisticated population models and relatively abundant and high-quality data (Maguire 1997).

Although numerous hypotheses have been proposed with respect to underlying causes, it is undoubtedly the case that limitations to model construction were substantial contributing factors (Myers *et al.* 1996). Erroneous assumptions in the virtual population analyses (VPA) conducted, based in part on poor understanding of spawning sites and migration pathways (Rose 1993), led to overestimation of population abundance, underestimation of fishing mortality and, consequently, the setting of quotas which were unsustainable. Unfortunately the errors in the models used, and the factors to which they had been, in effect, ignorant, were recognized too late to prevent collapse. Quite apart from the devastating social consequences of the closure of the fishery (Kurlansky 1998), it is likely that the collapse in stocks will have long-term impacts on genetic diversity and the maintenance of some migration routes (Rose 1993, Kulka *et al.* 1995).

9.4 THE PRECAUTIONARY PRINCIPLE AND ENVIRONMENTAL LEGISLATION

It was recognition of substantial and, to a degree, irreducible limitations to scientific knowledge which led initially to the formulation of a precautionary approach to environmental protection. The precautionary principle, or the principle of precautionary action, was developed to enable decisions on issues of environmental protection to be made on the basis of available evidence and 'incipient suspicion', but in the absence of demonstrated causality. This guiding principle had at its core the recognition that, in order to meet the responsibility of protecting the natural foundations of life for future generations, irreversible damage must be prevented by identifying and implementing protective measures in advance.

While the earliest origins of the *Vosorgeprinzip* remain unclear (DoE 1995, Gray and Bewers 1996, McIntyre and Mosedale 1997), its most complete definition is probably that given in a German Federal Government report on the protection of air quality (FRG 1986). This definition essentially comprises four elements:

1.   That damage should, as a priority, be avoided.
2.   That scientific research plays an essential role in identifying threats.
3.   That action to prevent harm is essential, even in absence of conclusive evidence of causality.
4.   That all technological developments should meet the requirement for progressive reduction of environmental burden.

Of these elements, the requirement for action in the absence of analytical or predictive certainty has become the most widely used condensation of the principle.

In essence, the precautionary principle stresses that action in anticipation of harm is essential to ensure that it does not occur (Bodansky 1991). The adoption of such an approach implies a shift in emphasis in favour of a bias towards environmental safety (McIntyre and Mosedale 1997), ensuring that any errors of judgement made will lead to excess, rather than inadequate, protection. In other words, the approach implies an acceptance that, in order to minimize the frequency of measures which are under-protective, some measures will inevitably be over-protective.

Since its initial formulation, the principle has gained increasing acceptance as a fundamental guiding paradigm, and now forms a central component of numerous national and international legislative frameworks on the protection of the environment. The principle has perhaps gained its highest profile within the field of marine environmental protection, especially with respect to inputs of hazardous chemicals. For example, although by no means its earliest use, the explicit inclusion of elements of the principle within the 1987 North Sea Ministerial Declaration (MINDEC 1987) represented a highly significant endorsement which, undoubtedly, subsequently facilitated its adoption by other regional and global marine fora. Notable among these are the OSPAR Convention (formerly the Oslo and Paris Conventions, protecting the North

East Atlantic, OSPAR 1992, 1998a), the Barcelona Convention (Mediterranean, UNEP 1996) and the London Convention on dumping of wastes at sea (LC 1972, 1996).

Nevertheless, the principle clearly also has application at the science–policy interface in relation to the release of biological agents (including genetically modified organisms), the management of resource exploitation (e.g. fisheries) and, indeed, any field in which human activity might have substantial, far-reaching or even irreversible impacts. Indeed, the precautionary principle is incorporated as a guiding paradigm within the 1987 Montreal Protocol regulating ozone-depleting substances, the 1992 Climate Change Convention, the Rio Declaration on Environment and Development (RioDEC 1992) and the 1995 United Nations Agreement on High Seas Fishing (UN 1995).

The degree to which the principle has been implemented in the form of effective programmes and measures varies greatly between agreements. A more comprehensive review than that given here is provided by McIntyre and Mosedale (1997). Perhaps the most transparent and definitive commitments to implementation relate to international agreements on the North Sea and within the OSPAR Convention.

9.4.1 NORTH SEA MINISTERIAL DECLARATIONS (1984–95)

The Ministerial Declaration which arose from the First International Conference on the Protection of the North Sea, held in Bremen in 1984, incorporated the statement that North Sea states must not wait for proof of harmful effects before taking action. This commitment was made more explicit in the Ministerial Declaration from the Second North Sea Conference (MINDEC 1987) with the recognition that, in order to safeguard the marine ecosystem, it would be necessary to reduce polluting emissions. Particular focus was placed on substances which possessed the hazard marker properties of persistence, toxicity and liability to bioaccumulate:

> "especially when there is reason to assume that certain damage or harmful effects on the living resources of the sea are likely to be caused by such substances, even when there is no scientific evidence to prove a causal link between emissions and effects (the principle of precautionary action)" (MINDEC 1987).

This interpretation was further strengthened by the third and fourth Ministerial Declarations (MINDEC 1990, 1995), the latter committing to continuous reduction of such inputs with a target of their cessation within one generation (i.e. by 2020). In the sense that this commitment addresses a very broad group of chemicals on the basis of their inherent hazardous properties, without the requirement for individual assessment of causality, it is truly precautionary in nature. Precisely how effectively such measures will translate to precautionary action remains to be seen.

9.4.2 THE OSPAR CONVENTION (1992)

The 1992 Convention for the Protection of the Marine Environment of the North East Atlantic (OSPAR 1992) adopted similar provisions, although the specific timeline for implementation (the 'one generation goal') was only adopted at the OSPAR Ministerial meeting in Sintra, Portugal, in July 1998 (OSPAR 1998a). Again, the commitment has been made to make every endeavour to achieve 'zero discharge' (cf. discharges, emissions and losses) for all hazardous substances to the OSPAR maritime region by 2020, with the aim of achieving concentrations close to zero in the environment for synthetic substances and close to background for naturally occurring substances. Hazardous substances are defined generally as those which are toxic, persistent and bioaccumulative or which give rise to an equivalent level of concern. In this regard, effects with long latency periods or which may impact only on subsequent generations, including carcinogenicity, teratogenicity and adverse effects on endocrine function, are explicitly recognized as forms of toxicity.

It is envisaged that implementation will be a staged process, with action on a priority list of chemicals of particular concern, including the brominated flame retardants (Box 9.2), within a more limited time-frame. Despite the fact that knowledge on fate and effects of these priority chemicals in the marine environment remains very limited, the Convention recognises that sufficient data exist to indicate that these are particularly hazardous with a strong likelihood that they may present substantial problems within the marine environment.

9.4.3 THE ROLE OF EUROPEAN COMMUNITY LEGISLATION

It must be recognized that priority listing in itself does not address the problems which these chemicals present. In this regard, there exists the potential for substantial conflict between the precautionary approach to chemical regulation adopted within OSPAR, for example, and that operating within the European Community, the latter based fundamentally on chemical-by-chemical assessment and management of risk.

The possibility remains, therefore, that precautionary measures developed under OSPAR will not be effectively implemented under European Community legislation. Further conflict arises as the Community is primarily an economic entity, such that the overriding commitments remain the economic and social development of the Community and its regions and the maintenance of the single European market. There is a strong adherence to the position that measures introduced to control the manufacture and use of chemicals must be harmonized across Europe and must not threaten to introduce barriers to trade. The need to maintain the competitiveness of Europe within the global market, increasingly stressed, promises to place further restrictions on the strength of measures designed to protect the environment even within regional fora.

The EC Treaty (EC 1993) summarizes the European Community approach to the environment under Article 130r. Interestingly, while the treaty bases environmental

---

Box 9.2   Case study: brominated flame retardants

It has long been recognized that brominated flame retardants, including poly-brominated biphenyls (PBBs) and polybrominated diphenyl ethers (PBDEs) are toxic, persistent and bioaccumulative (Cramer *et al.* 1990, Kamrin and Fischer 1991, Robertson *et al.* 1991, Jansson *et al.* 1993, Kholkute *et al.* 1994). Understanding of the breadth of toxic effects of these groups is developing rapidly (Hornung *et al.* 1996, Kang *et al.* 1996). Recently, for example, Eriksson *et al.* (1998) reported subtle impacts on brain development in rodents, resulting in permanent changes in behaviour, memory and learning, although the mechanisms have yet to be elucidated. It is also known that PBDEs can reduce the levels of circulating thyroid hormones in blood plasma (Darnerud and Sinjari 1996) and impact retinoid levels (Olsson *et al.* 1998).

Moreover, the compound tetrabromobisphenol-A (TBBPA) introduced into many products (including widespread use in computer hardware) as a substitute for other brominated flame retardants, may be equally hazardous (Sellstrom and Jansson 1995). Meerts *et al.* (1998) reported recently that TBBPA can compete with thyroxine for binding sites on human transthyretin protein.

These chemicals are clear targets for substitution on a precautionary basis as, although significant uncertainties remain, available evidence does indicate the potential for widespread and long-term effects resulting from the continued use and release of these chemicals. Some of these effects are likely to be serious or irreversible. Recent research which has confirmed the widespread presence of these contaminants in marine mammals and in human tissues (Klasson-Wehler *et al.* 1997, de Boer *et al.* 1998) merely strengthens the case for urgent protective measures.

It may be argued that any action taken to address this group of chemicals, while cautionary, would no longer be precautionary, as the chemicals are already widespread contaminants and, as such, may have caused significant damage. While this does not diminish the need for action, it may be hoped that precautionary measures may lead, in future, to avoidance of chemicals with similar properties, before they are manufactured, used and, consequently, released to the environment.

---

legislation on the precautionary principle, no further definitions are given and it remains unclear precisely how the precautionary principle may be implemented within or alongside existing instruments. There would, for example, appear to be little room for truly precautionary measures within existing directives controlling the marketing and use of chemicals. The absence of precaution is particularly apparent with respect to current permissive legislation on non-assessed 'existing chemicals', many of which are marketed in the absence of information on their potential environmental fates and effects (EEA 1998).

The processes by which the production, marketing and use of chemicals are regulated in the EU and in some individual Member States are currently undergoing review. The manner in which the European Commission intends to apply the precautionary principle was outlined in February 2000 (CEC 2000), though the suggested approach adheres strictly to risk-based assessment and management, with recourse to the principle only when such processes are deemed to have failed.

9.5 THE PRECAUTIONARY PRINCIPLE AND SCIENCE

Acceptance of the precautionary principle as a means for the development of protective measures has been limited, in part, by misconceptions regarding its role and intentions. A number of these are reviewed and challenged below.

9.5.1 'THE PRECAUTIONARY PRINCIPLE IS UNSCIENTIFIC'

It has been argued (e.g. Gray 1990, Bewers 1995) that the precautionary principle, in essence, lies outside the bounds of science as it promotes preventative action even in the absence of proof of causality. Risk-based approaches are commonly presented as the science-based alternative. Such views do not appear to recognise that the principle is founded on the use of comprehensive, co-ordinated research in order to guide precautionary action. The fundamental difference between risk and precautionary approaches is not that one uses science while the other does not, but simply the way in which scientific evidence is employed for decision-making at the science–policy interface. The precautionary approach is, to a degree, less prescriptive in its evaluation of the need for action, in that it does not rely on a need explicitly to define and quantify risks, but rather on the more general application of scientific research as a means for the early detection of dangers to human or wildlife health or to the environment as a whole. The commitments within the North Sea and OSPAR processes, for example, clearly have a firm basis in science, as they rely on scientific research to identify those properties and, thereby, substances or groups of substances which are of concern. Nevertheless, in the requirement to address all substances with those properties, the legislation is also clearly precautionary in nature. It is in this manner that science can continue to play a central role in the formulation and implementation of effective environmental legislation without the need for a reliance on quantitative forecasting of fates and effects.

9.5.2 'THE PRINCIPLE IS SIMPLY ONE OF MANY TOOLS WITHIN THE RISK ASSESSMENT PROCESS AND SHOULD ONLY BE INVOKED WHEN INSUFFICIENT DATA ARE AVAILABLE TO COMPLETE THE ASSESSMENT OF RISKS'

We have argued elsewhere that the precautionary principle is, in its own right, a crucial scientific tool to mitigate threats to the environment (Johnston and Simmonds 1990, 1991). Clearly it is not intended as a substitute for a scientific approach, but rather as an overarching principle to guide decision-making in the absence of analytical or predictive certainty. It provides a mechanism to compensate for inherent uncertainty and indeterminacy in natural systems and a central paradigm for responsible, timely and definitive preventative action.

In short, the precautionary principle cannot and should not be subsumed under a risk assessment mechanism, as is also currently implied within guidance for risk assessment and management in the UK (DoE 1995), and the Commission's communication (CEC 2000) to be invoked only when a risk assessment is judged to have failed (Brown 1998). Neither should risk assessment be seen as a means of implementing the precautionary principle, as one 'tool' in the full suite of risk assessment methodologies. Contrary to Brown (1998), the precautionary principle should operate at all times in recognition of the fact that assessments of hazards, exposure and risk, despite their apparent objectivity, can never alone ensure an adequate level of environmental protection. Indeed, if the principle is not operational at all times its effectiveness is greatly diminished.

9.5.3 'THE PRINCIPLE CAN BE REPLACED THROUGH INCORPORATION OF PESSIMISTIC ASSUMPTIONS WITHIN RISK ASSESSMENTS'

Gray and Bewers (1996) suggest that, in the context on the North Sea ministerial agreements, the precautionary principle should be implemented through the employment of pessimistic assumptions in standard risk assessment procedures. Such an approach captures neither the spirit nor the provisions intended for the principle and threatens to undermine its utility by subjecting it to the self same limitations of risk assessment and management procedures. Their arguments are challenged in more detail by Santillo *et al.* (1998).

9.5.4 'IMPLEMENTING THE PRECAUTIONARY PRINCIPLE WILL SIMPLY LEAD TO THE TRANSFERENCE OF IMPACTS FROM ONE MEDIUM TO ANOTHER'

For example, Bodansky (1991) argues that the choice faced by environmental regulators will always be between one risk and another and that the precautionary withdrawal of one process may simply lead to the transfer of the problem to other media. This view relies very much on the assumption that the precautionary principle will be implemented in a simple one-sided, approach to decision-making, without consideration of the potential hazards of alternatives. In contrast, for action to be truly precautionary, it must also ensure that the fundamental objective of the reduction of overall environmental burden is strictly observed. In some legislative fora, this requirement is made explicit. For example, the 1996 Protocol to the London Convention (LC 1996), although relating fundamentally to the dumping of wastes at sea, nevertheless contains a clear commitment to address all sources of pollution.

In order to meet the objective of comprehensive environmental protection, it must be recognized that a decision, for example, to prevent the use or discharge of a certain chemical may require a fundamental re-evaluation of societal need for that product and may not always imply simple substitution with an alternative. For example, if the commitments within the North Sea ministerial process (MINDEC 1995) and, more recently, under OSPAR (1998a), are to be met, particularly to achieve zero discharge, emissions and losses of hazardous substances to the

North Sea and North-east Atlantic regions respectively, changes to industrial practice, process and even products will undoubtedly be necessary.

9.6 CONCLUSIONS

The process of arriving at decisions on issues of environmental protection, and particularly the development of measures to prevent degradation, will always necessitate an element of predictive analysis. While both the risk-based and precautionary approaches outlined above rely on the provision of high-quality scientific data to guide decision-making, they differ fundamentally in the manner in which these data are employed. Their underlying philosophies are clearly very different, as are the assumptions on which they are founded and the extent to which each makes allowances for the existence of uncertainties, indeterminacies and unidentified pathways and effects.

Risk-based methods employ scientific knowledge in a prescriptive manner, in order to arrive at quantitative forecasts of fate and effects. This approach implies an acceptance that intrinsic hazards associated with a chemical do not, in themselves, present problems providing that exposure can be adequately managed. Notwithstanding the potential for the use of pessimistic assumptions within the assessments, such forecasts are heavily optimistic in their assumptions that all or, at least, the most sensitive pathways and end points have been considered and that chemicals can be managed in such a way that exposure may be accurately described and controlled.

In contrast, implementation of the precautionary principle involves the use of scientific knowledge less prescriptively, as a guide for the development of protective measures. This latter approach recognizes that accurate prediction of precise fates and effects of chemicals is likely to remain an unachievable goal and is not a sound basis on which to develop policies for environmental protection. Within the OSPAR *Strategy for Hazardous Substances* (OSPAR, 1998b), for example, the intrinsic properties of persistence, toxicity and bioaccumulative capacity are the primary considerations when selecting chemicals for further action, without the need for demonstration of widespread occurrence or effects in the marine environment.

Clearly, further scientific research and methods development are essential if we are to elucidate mechanisms and become more effective at identifying and predicting environmental threats. Nevertheless, scientific research can only guide policy-making; it cannot decide policy in itself. Neither can it predict with absolute certainty future trends and effects. In issues of environmental protection, particularly where threats of serious or irreversible damage are identified, a decision to defer judgement until such time as further data become available cannot be a responsible option. If the need for action is retrospectively identified, some damage is likely already to have occurred.

The North Sea ministerial process and the OSPAR Convention provide perhaps some of the strongest bases yet for the practical implementation of the precautionary principle as it applies to the protection of the marine environment, particularly with regard to hazardous substances. It is now essential to ensure that the principle is strictly observed during the further development of these agreements, and particularly during the development and application of practical programmes and measures to address threats to the marine environment. Moreover, it is important that similar levels of protection are conferred to other compartments of the environment. It is only through adopting mechanisms which enable and, indeed, require precautionary action that we will be capable of ensuring that environmental damage and threats to human health may, wherever possible, be avoided in advance.

Professor James Utterback (Sherden 1998) concluded that: 'The illusion of knowing what's going to happen is worse than not knowing'. While seeming to present an entirely negative view, his conclusion nevertheless serves as a reminder of the fundamental limitations to the application of predictive science in helping us prepare for and control the future. This is not to say that science should not strive to fill gaps in knowledge and understanding, but merely that we must remain aware of the limitations to that understanding and make allowances for unexpected pathways and events in decisions regarding the future.

REFERENCES

Berg M and Scheringer M (1994) Problems in environmental risk assessment and the need for proxy measures. *Fresenius Environment Bulletin*, **3**, 487–492.
Bewers JM (1995) The declining influence of science on marine environmental policy. *Chemistry and Ecology*, **10**, 9–23.
Bodansky D (1991) Scientific uncertainty and the Precautionary Principle. *Environment*, **33**(7), 4–5 and 43–44.
Brown D (1998) Environmental risk assessment and management of chemicals. In *Issues in Environmental Science and Technology 9: Risk Assessment and Risk Management*, Hester RE and Harrison RM (eds), The Royal Society of Chemistry, Cambridge, UK, pp. 91–111.
Burger J and Gochfeld M (1997) Paradigms for ecological risk assessment. In *Preventative Strategies for Living in a Chemical World*, Bingham E and Rall DP (eds). *Annals of the New York Academy of Sciences*, **837**, 372–386.
CEC (2000) *Communication from the Commission on the precautionary Principle*, CEC, Brussels, 2.2.2000, COM (2000-1 Final), 28 pp.
Cramer PH, Ayling RE, Thornburg KR, Stanley JS, Remmers JC, Breen JJ and Schwemberger J (1990) Evaluation of an analytical method for the determination of polybrominated dibenzo-p-dioxins/dibenzofurans (PBDD/PBDF) in human adipose. *Chemosphere*, **20**(7–9), 821–827.
Darnerud PO and Sinjari T (1996) Effects of polybrominated diphenyl ethers (PBDEs) and polychlorinated biphenyls (PCBs) on thyroxine and TSH blood levels in rats and mice. *Organohalogen Compounds*, **29**, 316–319.
De Boer J, Wester PG, Klamer HJC, Lewis WE and Boon JP (1998) Do flame retardants threaten ocean life? *Nature*, **394**, 28–29.
DoE (1995) *A Guide to Risk Assessment and Risk Management for Environmental Protection*. Department of the Environment, UK, 92 pp.
Dovers SR and Handmer JW (1995) Ignorance, the precautionary principle and sustainability. *Ambio*, **24**(2), 92–97.
Ducrotoy J-P and Elliott M (1997) Interrelations between science and policy-making: the North Sea example. *Marine Pollution Bulletin*, **34**, 686–701.
EC (1993) *Treaty Establishing the European Community*, Article 130r: 297–299.

EEA (1998) *Chemicals in the European Environment: Low Doses, High Stakes?* The EEA and UNEP Annual Message 2 on the State of Europe's Environment, European Environment Agency, Copenhagen, 32 pp.

Eriksson P, Jakobsson E and Fredriksson A (1998) Developmental neurotoxicity of brominated flame retardants: polybrominated diphenyl ethers and tetrabromobiosphenol-A. *Organohalogen Compounds*, **35**, 375–377.

FRG (1986) *Umweltpolitik: Guidelines on Anticipatory Environmental Protection.* Federal Ministry for the Environment, Nature Conservation and Nuclear Safety, Germany, 43 pp.

Funtowicz SO and Ravetz JR (1994) Uncertainty, complexity and post-normal science. *Environmental Toxicology and Chemistry*, **13**(12), 1881–1885.

Gray JS (1990) Statistics and the precautionary principle. *Marine Pollution Bulletin*, **21**, 174–176.

Gray JS and Brewers JM (1996) Towards a scientific definition of the precautionary principle. *Marine Pollution Bulletin*, **32**(11), 768–771.

Hornung MW, Zabel EW and Peterson RE (1996) Toxic equivalency factors of polybrominated dibenzo-*p*-dioxin, dibenzofuran, biphenyl, and polyhalogenated diphenyl ether congeners based on rainbow trout early life stage mortality. *Toxicology and Applied Pharmacology*, **140**(2), 227–234.

Hutchings JA (1996) Spatial and temporal variation in the density of northern cod and a review of hypotheses for the stocks collapse. *Canadian Journal of Fisheries and Aquatic Sciences*, **53**(5), 943–962.

Hutchings JA and Myers RA (1994) What can be learned from the collapse of a renewable resource? Atlantic cod (*Gadus morhua*) off Newfoundland and Labrador. *Canadian Journal of Fisheries and Aquatic Sciences*, **51**, 2126–2146.

Jansson B, Andersson R, Asplund L, Litzen K, Nylund K and Sellstrom U (1993) Chlorinated and brominated persistent organic-compounds in biological samples from the environment. *Environmental Toxicology and Chemistry*, **12**(7), 1163–1174.

Johnston P and Simmonds M (1990) Precautionary principle (letter). *Marine Pollution Bulletin*, **21**, 402.

Johnston P and Simmonds M (1991) Green light for precautionary science. *New Scientist*, **3 August**, 14.

Johnston PA, Santillo D and Stringer RL (1996) Risk assessment and reality: Recognising the limitations. In *Environmental Impact of Chemicals: Assessment and Control.* Quint M, Purchase R and Taylor D (eds), Royal Society of Chemistry, Cambridge, ISBN 0-85404-795-6, 223–239.

Kamrin MA and Fischer LJ (1991) Workshop on human health impacts of halogenated biphenyls and related compounds. *Environmental Health Perspectives*, **91**, 157–164.

Kang KS, Wilson MR, Hayashi T, Chang CC and Trosko JE (1996) Inhibition of gap functional intercellular communication in normal human breast epithelial cells after treatment with pesticides, PCBs and PBBs, alone or in mixtures. *Environmental Health Perspectives*, **104**(2), 192–200.

Kholkute SD, Rodriguez J and Dukelow WR (1994) The effects of polybrominated biphenyls and perchlorinated terphenyls on *in-vitro* fertilization in the mouse. *Archives of Environmental Contamination and Toxicology*, **26**(2), 208–211.

Klasson-Wehler E, Hovander L and Bergman A (1997) New organohalogens in human plasma — identification and quantification. *Organohalogen Compounds*, **33**, 420–425.

Kulka DW, Wroblewski JS and Narayanan S (1995) Recent changes in the winter distribution and movements of Northern Atlantic cod (*Gadus morhua* Linnaeus, 1758) on the Newfoundland–Labrador Shelf. *ICES Journal of Marine Science*, **52**, 889–902.

Kurlansky M (1998) *Cod: A Biography of the Fish that Changed the World*, Jonathan Cape, London, 294 pp.

LC (1972) *Convention on the Prevention of Marine Pollution by Dumping of Wastes and Other Matter, 1972. The London Convention*, LC.2/Circ. 380. International Maritime Organization, London, 63 pp.

LC (1996) *1996 Protocol to the Convention on the Prevention of Marine Pollution by Dumping of Wastes and Other Matter*, LC.2/Circ. 380, International Maritime Organization, London, 63 pp.

Maguire J-J (1997) The precautionary approach to fisheries: lessons from the collapse of the Canadian groundfish stocks. *North Sea Monitor*, **15**(3), 4–8.

McIntyre O and Mosedale T (1997) The precautionary principle as a norm of customary international law. *Journal of Environmental Law*, **9**(2), 221–241.

Meerts IATM, Marsh G, van Leeuwen-Bol I, Luijks EAC, Jakobsson E, Bergman A and Brouwer A (1998) Interaction of polybrominated diphenyl ether metabolites (PBDE–OH) with human transthyretin *in vitro*. *Organohalogen Compounds*, **37**, 309–312.

Meironyte D, Bergman A and Noren K (1998) Analysis of polybrominated diphenyl ethers in human milk. *Organohalogen Compounds*, **35**, 387–388.

MINDEC (1987) *Ministerial Declaration of the Second International Conference on the Protection of the North Sea*, 24–25 November 1987, London, UK.

MINDEC (1990) *Ministerial Declaration of the Third International Conference on the Protection of the North Sea*, 7–8 March 1990, The Hague, The Netherlands.

MINDEC (1995) *Ministerial Declaration of the Fourth International Conference on the Protection of the North Sea*, 8–9 June 1995 Esbjerg, Denmark.

Myers RA, Hutchings JA and Barrowman NJ (1996) Hypotheses for the decline of cod in the North Atlantic. *Marine Ecology Progress Series*, **138**, 293–308.

Olsson P-E, Borg B, Brunstrom B, Hakansson H and Klasson-Wehler E (1998) *Endocrine Disrupting Substances—Impairment of Reproduction and Development*, Swedish Environmental Protection Agency Report 4859, ISBN 91-620-4859-7, 150 pp.

OSPAR (1992) *Final Declaration of the Ministerial Meetings of the Oslo and Paris Commissions*. Oslo and Paris Conventions for the Prevention of Marine Pollution, Paris 21–22 September.

OSPAR (1998a) *The Sintra Statement (Final Declaration of the Ministerial Meeting of the OSPAR Commission, Sintra 20–24th July 1998)*. OSPAR 98/14/1 Annex 45. OSPAR Convention for the Protection of the Marine Environment of the North-East Atlantic.

OSPAR (1998b) *OSPAR Strategy with Regard to Hazardous Substances*. OSPAR 98/14/1 Annex 34. OSPAR Convention for the Protection of the Marine Environment of the North-East Atlantic.

Power M and McCarty LS (1997) Fallacies in ecological risk assessment practices. *Environmental Science and Technology*, **31**, 370A–375A.

Rescher N (1998) *Predicting the Future: An Introduction to the Theory of Forecasting*, State University of New York Press, ISBN 0-7914-3553-9, 315 pp.

RioDEC (1992) *Rio Declaration on Environment and Development*, ISBN 9-21-100509-4, 1992.

Robertson LW, Silberhorn EM, Glauert HP, Schwarz M and Buchmann A (1991) Do structure–activity-relationships for the acute toxicity of PCBs and PBBs also apply for induction of hepatocellular-carcinoma? *Environmental Toxicology and Chemistry*, **10**(6), 715–726.

Rose GA (1993) Cod spawning on a migration highway in the North West Atlantic. *Nature*, **366**, 458–461.

Santillo D, Stringer RL, Johnston PA and Tickner J (1998) The precautionary principle: Protecting against failures of scientific method and risk assessment. *Marine Pollution Bulletin*, **36**(12), 939–950.

Santillo D, Johnston P and Stringer RL (2000) Management of chemical exposure: the limitations of a risk-based approach. *International Journal of Risk Assessment and Management*, **1**, 160–180.

Sellstrom U and Jansson B (1995) Analysis of tetrabromobisphenol A in a product and environmental samples. *Chemosphere*, **31**(4), 3085–3092.

Sherden WA (1998) *The Fortune Sellers: The Big Business of Buying and Selling Predictions*, Wiley, New York, ISBN 0-471-18178-1, 308 pp.

UN (1995) *Agreement for the Implementation of the Provisions of the United Nations Convention on the Law of the Sea of 10 December 1982 Relating to the Conservation and Management of Straddling Fish Stocks and Highly Migratory Fish Stocks, The Earth Summit Agreement on High Seas Fishing*. United Nations, NY, November 1995, 43 pp.

UNEP (1996) *Final Act of the Conference of the Plenipotentiaries on the Amendments to the Convention for the Protection of the Mediterranean Sea Against Pollution from Land-Based Sources*. UNEP(OCA)/MED 7/4, United Nations Environment Programme, March 1996.

Wynne B (1992) Uncertainty and environmental learning: reconceiving science and policy in the preventative paradigm. *Global Environmental Change*, **June**, 111–1127.

10

Modelling of Toxicity to the Ciliate *Tetrahymena pyriformis*: the Aliphatic Carbonyl Domain

MARK T. D. CRONIN[1], GLENDON D. SINKS[2] AND T. WAYNE SCHULTZ[2,3]

[1]*School of Pharmacy and Chemistry, Liverpool John Moores University, Liverpool, UK*
[2]*College of Veterinary Medicine, The University of Tennessee, Knoxville TN, USA*
[3]*Waste Management Research and Education Institute, The University of Tennessee, Knoxville TN, USA*

10.1 INTRODUCTION

Quantitative structure–activity relationships (QSARs) are predictive techniques that attempt to model biological activity (Martin 1978) and therefore have an important role in attempts to forecast the potential toxicological effects of a chemical. A QSAR may be defined as an attempt to relate statistically the biological activity (or, for the purposes of this chapter, toxicological activity) of a series of chemicals to some feature of their physicochemical, or structural properties (Cronin 1998). Implicit in the statistical modelling is that the goodness of fit of the model is assessed, and some form of validation is undertaken. In the field of the environmental sciences, QSARs may be used to forecast a number of effects such as toxicity, penetration through membranes, bioaccumulation, biodegradation and metabolism (Karcher and Devillers 1990).

It is of immense concern to consider that of the 70 000 chemicals in daily use, there are reliable data for fate processes and effect concentrations for a very limited number of these compounds (Hermens 1991). The advantage of QSAR as a predictive technique to forecast the toxicity of chemicals is thus apparent. It potentially provides a technique for forecasting (and so prioritizing for testing) the toxicological effects of existing chemicals, as well as providing a technique for the reduction of animal tests on new chemicals. Other advantages of QSAR include some ability to shed light on the mechanism of toxic action for a series of compounds, and to highlight inter-species variabilities and sensitivities.

An appreciation of the mechanism of toxic action is of primary importance for the development of QSARs for toxicity (Cronin and Dearden 1995). Traditionally, QSARs have been developed for single mechanisms of action. The identification

Forecasting the Environmental Fate and Effects of Chemicals. Edited by Philip S. Rainbow, Steve P. Hopkin and Mark Crane.
© 2001 John Wiley & Sons Ltd

of a toxic mechanism of action is seldom easy (Schultz et al. 1997), and workers have been restricted to the development of QSARs for 'structurally similar' compounds, or congeneric series, e.g. phenols and nitrobenzenes (Cronin and Dearden 1995). It is assumed that compounds with the same functional group, i.e. the aromatic hydroxy or nitro group, are acting by the same mechanism of action. While the mechanism of action may remain the same for a modest group of substituents, even a modicum of structural heterogeneity (e.g. a single chlorine on a nitrobenzene ring) may alter the toxic mechanism of action. This issue is exacerbated when mixtures of classes are considered, for example is 4-hydroxynitrobenzene, a phenol, or a nitrobenzene?

In modelling toxicity, two main modes of toxicity are apparent (for the purposes of this discussion receptor-based toxicities are ignored) (Karabunarliev et al. 1996). These are characterized as the narcotic and electrophilic modes of action. The narcoses mechanisms of action are the most common, accounting for the toxicity of up to 70% of organic chemicals (Bradbury and Lipnick 1990). Narcosis is a non-reactive toxicity characterized by the ability to disrupt membrane function (van Wezel and Opperhuizen 1995). Generally such toxicity can be modelled by hydrophobicity (Könemann 1981). Electrophilic toxicants, which account for the toxicity of the majority of the remaining compounds, are 'reactive' and have the ability to form covalent bonds with nucleophilic sites on macromolecules. To model successfully the toxicity of electrophilic compounds, parameters that describe electrophilic reactivity must be employed (Mekenyan and Veith 1994).

Thus to model toxicity a generic mechanism, which consists of two processes, is considered. The first process is the ability of the xenobiotic to pass through, or enter into, a phospholipid bilayer or membrane. (This membrane may be as varied as a fish gill or stomach wall, human skin, or a plasmalemma). This process is modelled by hydrophobicity (Könemann 1981). The second process is one of interaction of the xenobiotic with a biological macromolecule such as a protein or DNA. This process is modelled by electronic parameters (Mekenyan and Veith 1994). A third, and lesser, process, which may often be modelled by electronic factors, is the role of metabolism. To model such a two-process effect, McFarland (1970) demonstrated the use of a two-parameter QSAR model:

$$\log 1/C = a \log \text{(penetration)} + b \log \text{(interaction)} + c \tag{1}$$

where C is the concentration causing the response, a and b are the coefficients, or slopes, for each variable and c a constant, or the intercept.

Veith and Mekenyan (1993) demonstrated, for the modelling of toxicity, the use of the hydrophobic term the logarithm of the 1-octanol/water partition coefficient (log K_{OW}) to model penetration and the electronic term the energy of the lowest unoccupied molecular orbital (E_{LUMO}) to model molecular interaction.

$$\log 1/\text{toxicity} = a \log K_{OW} + b \, E_{LUMO} + c \tag{2}$$

Veith and Mekenyan (1993) reported this approach to be successful in the modelling of the toxicity of a variety of chemicals to the fish *Pimephales promelas*. The applicability of the two-parameter QSAR has also been demonstrated in modelling the toxicity of aliphatic compounds to the bacterium *Vibrio fischeri* (Cronin and Schultz 1998). Recently, Cronin *et al.* (1998) demonstrated the use of such a two-parameter QSAR, or response-surface approach to model the toxicity of 42 alkyl and halogen substituted mono- and di-nitrobenzenes to *Tetrahymena pyriformis* (log (IGC_{50}^{-1})). These nitrobenzenes were known to have a number of different mechanisms of toxic action including polar narcosis, pro-electrophilic activity and $S_N Ar$ electrophilic reactivity. The response-surface developed was:

$$\log(IGC_{50}^{-1}) = 0.47 \log K_{OW} - 0.60 \, E_{LUMO} - 2.55$$
$$n = 42, \ r^2 = 0.881, \ s = 0.246, \ F = 154, \ Q^2 = 0.866 \tag{3}$$

where: n is the number of observations, r^2 the coefficient of determination, s the standard error of the estimate, F Fisher's statistic and Q^2 the leave-one-out cross-validated coefficient of determination.

The aim of this chapter, therefore, is to illustrate that QSARs have considerable potential to predict the toxicity of chemicals to environmentally important species, these predictions being based upon a knowledge of physico-chemical structure alone. To this end, the chapter extends the work of Schultz and Cronin (1999) and Schultz and Deweese (1999) on the aliphatic 'domain' of the *Tetrahymena* toxicity response surface by the modelling of the toxicity of a series of 'structurally similar' aliphatic compounds. These compounds are 'similar' in that they all contain a carbonyl group (C=O). In the different molecular configurations examined, there appear to be a number of different mechanisms of toxic action, not all of which are currently known.

10.2 TOXICITY OF ALIPHATIC COMPOUNDS TO *TETRAHYMENA PYRIFORMIS*

10.2.1 COMPOUNDS TESTED

The toxicity of the 140 aliphatic carbonyl-containing compounds has been obtained experimentally (see Table 10.1). Despite the apparent chemical simplicity of these compounds they form 12 distinct chemical groups with a variety of putative mechanisms of action (see Table 10.2). Ketones and aldehydes with a double bond in the alkyl chain but not in the α-position (e.g. 5-methyl-2-hexanone; (*cis*)-7-decen-1-al), are considered to be acting as saturated ketones and aldehydes respectively.

10.2.2 *TETRAHYMENA PYRIFORMIS* TOXICITY TEST

Population growth impairment testing with the common ciliate, *Tetrahymena pyriformis* (strain GL-C) was conducted following the protocol described by

Table 10.1 The name, toxicity, hydrophobicity and electrophilicity of the compounds considered in this study

Chemical	CAS number	$\log(\text{IGC}_{50}^{-1})$	$\log K_{OW}$	E_{LUMO}
Ketones				
acetone	67-64-1	−2.15	−0.24	0.8349
2-butanone	78-93-3	−1.75	0.29	0.8796
3-pentanone	96-22-0	−1.46	0.82	0.9121
2-hexanone	591-78-6	−1.34	1.38	0.8793
2-pentanone	107-87-9	−1.22	0.84	0.8767
4-methyl-2-pentanone	108-10-1	−1.21	1.31	0.8980
3-methyl-2-butanone	563-80-4	−1.17	0.56	0.9093
5-hexen-2-one	109-49-9	−1.14	1.02	0.8210
5-methyl-5-hexen-2-one	3240-09-3	−0.87	1.29	0.9163
4-heptanone	123-19-3	−0.67	1.91	0.9039
5-methyl-2-hexanone	110-12-3	−0.65	1.88	0.8897
2-heptanone	110-43-0	−0.49	1.98	0.8699
6-methyl-5-hepten-2-one	110-93-0	−0.45	2.21	0.8638
2-octanone	111-13-7	−0.15	2.37	0.8760
5-nonanone	502-56-7	0.07	2.97	0.9145
2-decanone	693-54-9	0.58	3.77	0.8793
3-decanone	928-80-3	0.63	3.49	0.9120
2-nonanone	821-55-6	0.66	3.16	0.8781
2-undecanone	112-12-9	1.50	4.09	0.8686
7-tridecanone	462-18-0	1.52	5.08	0.9122
2-dodecanone	6175-49-1	1.67	4.55	0.8724
2-tridecanone	593-08-8	2.12	5.08	0.8762
Mono-esters				
methyl acetate	79-20-9	−2.16	0.18	1.0980
ethyl acetate	141-78-6	−1.78	0.73	1.1487
methyl propionate	554-12-1	−1.62	0.82	1.1484
isopropyl acetate	108-21-4	−1.59	1.02	1.1896
(tert)-butyl acetate	540-88-5	−1.49	1.76	1.2556
methyl trimethylacetate	598-98-1	−1.36	1.83	1.2350
isobutyl acetate	110-19-0	−1.27	1.78	1.1526
ethyl isobutyrate	97-62-1	−1.27	1.55	1.2486
methyl butyrate	623-42-7	−1.26	1.29	1.1522
propyl acetate	109-60-4	−1.24	1.24	1.1494
(+/−)-methyl-2-methylbutyrate	53955-81-0	−1.17	1.55	1.2062
ethyl propionate	105-37-3	−0.95	1.21	1.1974
(+/−)-ethyl-2-methylbutyrate	7452-79-1	−0.89	1.59	0.7058
methyl valerate	624-24-8	−0.84	1.96	1.1535
propyl propionate	106-36-5	−0.81	1.77	1.1981
ethyl isovalerate	108-64-5	−0.72	2.17	1.2135
isobutyl propionate	540-42-1	−0.69	2.17	1.2006
(+/−)-(sec)-butyl acetate	105-46-4	−0.68	1.72	1.1956
isobutyl isobutyrate	97-85-8	−0.67	2.48	1.2443
methyl hexanoate	106-70-7	−0.56	2.30	1.1527
butyl acetate	123-86-4	−0.49	1.78	1.1489
ethyl butyrate	105-54-4	−0.49	1.77	1.2009
(*tert*)-butyl propionate	20487-40-5	−0.41	1.95	1.3011
propyl butyrate	105-66-8	−0.41	2.30	1.1979
ethyl valerate	539-82-2	−0.36	2.30	1.2018
2-ethylbutyl acetate	1003-87-5	−0.12	2.70	1.1598

Continued

Table 10.1   *Continued*

Chemical	CAS number	$\log(\mathrm{IGC}_{50}^{-1})$	$\log K_{\mathrm{OW}}$	E_{LUMO}
amyl propionate	624-54-4	−0.04	2.83	1.2031
propyl valerate	141-06-0	0.01	2.83	1.1973
ethyl hexanoate	123-66-0	0.06	2.83	1.2012
methyl heptanoate	106-73-0	0.10	2.83	1.1529
amyl acetate	628-63-7	0.16	2.30	1.1491
butyl propionate	590-01-2	0.17	2.30	1.2016
butyl butyrate	109-21-7	0.52	2.83	1.2013
methyl octanoate	111-11-5	0.54	3.36	1.1522
methyl nonanoate	1731-84-6	0.96	3.88	1.1520
octyl acetate	112-14-1	1.06	3.88	1.1480
methyl decanoate	110-42-9	1.38	4.41	1.1515
methyl undecanoate	1731-86-8	1.42	4.94	1.1523
decyl acetate	121-17-4	1.88	4.94	1.1475
Formates				
ethyl formate	109-94-4	−1.63	0.26	1.1328
methyl formate	107-31-3	−1.62	0.03	1.0761
propyl formate	110-74-7	−1.46	0.83	1.1359
(*tert*)-butyl formate	762-75-4	−1.37	0.97	1.2760
isobutyl formate	542-55-2	−1.33	1.19	1.1392
butyl formate	592-84-7	−0.93	1.32	1.3668
n-amyl formate	638-49-3	−0.78	1.85	1.1374
n-hexyl formate	629-33-4	−0.38	2.38	1.1368
Aldehydes				
propionaldehyde	123-38-6	−0.65	0.59	0.8663
butyraldehyde	123-72-8	−0.55	0.88	0.8746
2-methylvaleraldehyde	123-15-9	−0.47	1.67	0.9009
isobutyraldehyde	78-84-2	−0.43	0.61	0.8970
2-methylbuteraldehyde	96-17-3	−0.39	1.14	0.8686
3,3-dimethylbutyraldehyde	2987-16-8	−0.37	1.63	0.9199
isovaleraldehyde	590-86-3	−0.33	1.23	0.8762
hexanal	66-25-1	−0.17	1.78	0.8689
valeraldehyde	110-62-3	−0.13	1.36	0.8697
2-ethylbutyraldehyde	97-96-1	−0.05	1.67	0.9166
heptaldehyde	111-71-7	−0.002	2.42	0.8668
2-ethylhexanal	123-05-7	0.16	2.73	0.9142
octylaldehyde	124-13-0	0.45	2.95	0.8663
nonylaldehyde	124-19-6	0.81	3.48	0.8665
(*cis*)-7-decen-1-al	21661-97-2	0.95	3.52	0.8565
(*trans*)-4-decen-1-al	65405-70-1	1.21	4.05	0.8233
decylaldehyde	112-31-2	1.28	4.01	0.8668
undecylicaldehyde	112-44-7	1.60	4.54	0.8649
dodecylaldehyde	112-54-9	1.76	5.07	0.8656
α,β-*Unsaturated ketones*				
4-methyl-3-penten-2-one	141-79-7	−0.64	1.05	0.0585
3-penten-2-one	625-33-2	0.54	0.52	0.0542
3-hepten-2-one	1119-44-4	0.70	1.57	0.0669
3-octen-2-one	1669-44-9	0.74	2.23	0.0692
4-hexen-3-one	2497-21-4	0.93	1.78	0.0782
3-nonen-2-one	14309-57-0	0.98	2.63	0.0708
3-buten-2-one	78-94-4	1.51	0.12	0.0503

Continued

Table 10.1   *Continued*

Chemical	CAS number	$\log(\text{IGC}_{50}^{-1})$	$\log K_{\text{OW}}$	E_{LUMO}
α,β-Unsaturated esters				
(*trans*)-ethyl crotonate	623-70-1	−0.76	1.86	0.0246
isopropenyl acetate	108-22-5	−0.69	1.07	0.7086
methyl-2-hexenoate	13894-63-8	0.51	2.38	0.0193
methyl-(*trans*)2-octenoate	7367-81-9	0.76	3.44	0.0219
methyl-2-nonenoate	111-79-5	1.04	3.97	0.0190
α,β-Unsaturated aldehydes				
(*trans*)-2-methyl-2-butenal	497-03-0	−0.14	1.01	−0.1144
2-methyl-2-pentenal	623-36-9	−0.02	1.53	−0.0930
2,4-dimethyl-2,6-heptadienal	unknown	0.08	2.33	−0.1014
3-methyl-2-butenal	107-86-8	0.09	1.01	−0.1612
7-(*trans*)-2-pentenal	1576-87-0	0.66	1.05	−0.1179
2-butenal	123-73-9	0.70	0.52	−0.1411
2,4-hexadienal	142-83-6	0.75	1.15	−0.5950
(*trans*)-2-hexenal	6728-26-3	0.76	1.58	−0.1151
(*trans,trans*)-2,4-heptadienal	4313-03-5	0.86	1.68	−0.5831
(*trans*)-2-heptenal	18829-55-5	1.05	2.11	−0.1143
(*trans*)-2-octenal	2548-87-0	1.20	2.64	−0.1145
(*trans,trans*)-2,4-nonadienal	5910-87-2	1.22	2.74	−0.5816
(*trans*)-2-(*cis*)-6-nonadienal	557-48-2	1.34	2.68	−0.1256
acrolein	107-02-8	1.41	−0.01	−0.1389
(*trans*)-2-nonenal	18829-56-6	1.60	3.16	−0.1144
(*trans*)-2-decen-1-al	3913-81-3	1.85	3.69	−0.1151
α-Diones				
2,3-butanedione	431-03-8	−0.23	−1.34	−0.5171
2,3-hexanedione	3848-24-6	−0.21	−0.31	−0.4820
2,3-pentanedione	600-14-6	−0.16	−0.84	−0.4860
3,4-hexanedione	4437-51-8	−0.01	−0.31	−0.4775
2,3-heptanedione	96-04-8	0.04	0.22	−0.4805
β-Diones				
3,5-heptanedione	7424-54-6	−0.38	0.60	0.3805
2,4-pentanedione	123-54-6	−0.27	0.24	0.3971
2,4-octanedione	14090-87	0.13	1.13	0.4742
2,4-nonanedione	6175-23-1	0.51	1.65	0.4830
γ-Diones				
2,5-hexanedione	110-13-4	−1.40	−0.27	0.6633
Cyclic-Diones				
1,4-cyclohexanedione	637-88-7	−1.67	−0.52	0.3623
1,3-cyclohexanedione	504-02-9	−0.93	−0.45	0.3844
1,2-cyclohexanedione	765-87-7	−0.92	−0.85	−0.1083
Lactones				
γ-butyrolactone	96-48-0	−1.72	−0.64	1.1041
γ-valerolactone	108-29-2	−1.67	−0.28	1.1590
ε−caprolactone	502-44-3	−1.26	0.87	0.8672
γ-caprolactone	695-06-7	−1.24	0.24	1.1633
α-methyl-γ-butyrolactone	1679-47-6	−1.19	−0.28	1.1573
(+/−)-β-butyrolactone	36536-46-6	−0.76	−0.84	0.9889
γ-octanoic lactone	104-50-7	−0.38	1.30	1.1633
β-propiolactone	57-57-8	−0.13	−1.36	0.9133
(+/−)-δ-decanolactone	705-86-2	−0.08	2.39	1.2051
γ-nonanoic lactone	104-61-0	0.01	1.83	1.1628
γ-decanolactone	706-14-9	0.49	2.72	1.1622

Table 10.2 Chemical classes considered in this study and their putative mechanisms of toxic action

Chemical class	Mechanism of toxic action	Reference
Narcotic mechanism of action		
ketones	Non-polar narcosis	Veith *et al.* 1983
mono-esters	Non-polar narcosis	Jaworska *et al.* 1995
Electrophilic mechanisms of action		
aldehydes	Schiff's base formers	Hermens 1990
formates	Schiff's base formers	Hermens 1990
α,β-unsaturated ketones	Michael-type acceptors	Lipnick 1991
α,β-unsaturated esters	Michael-type acceptors	Lipnick 1991
α,β-unsaturated aldehydes	Michael-type acceptors	Lipnick 1991
α-diones	Selective binders to arginine	Goldschmidt and Lockhart 1971
γ-diones	Selective binders to tubulin	Cohen *et al.* 1997
lactones	Strained ring electrophiles	Hermens 1990
β-diones	Unknown[a]	—
cyclic-diones	Unknown[a]	—

[a]While the mechanisms of action of β- and cyclic-diones are currently unknown, these compounds exhibit toxicity in excess of non-polar narcosis, and due to their structural similarity to α- and γ-diones are presumed to be electrophilic in nature.

Schultz (1997). This 40 h assay is static in design and uses population density quantitated spectrophotometrically at 540 nm as its end point. The test protocol allows for eight or nine cell cycles in controls. Each chemical was tested for three replicates following range-finding. Two controls, one that had no test material, but had been inoculated with *T. pyriformis*, and a blank, which had neither test material nor ciliates, were used to provide a measure of the acceptability of the test by indicating the suitability of the medium and test conditions as well as a basis for interpreting data from other treatments. Each test replicate consists of six to eight different concentrations of each test material with duplicate flasks of each concentration. Only replicates with control-absorbency values >0.6, but <0.75 were used in the analyses.

The 50% growth inhibitory concentration, IGC_{50}, was determined for each compound by Probit Analysis of Statistical Analysis System (SAS) software (SAS Institute Inc. 1989) with Y as the absorbency normalized as percentage of control and X as the toxicant concentration in $mg\,l^{-1}$. IGC_{50} values were converted to millimolar units for QSAR analysis.

10.2.3 CALCULATION OF PHYSICO-CHEMICAL PROPERTIES

Values for the hydrophobicity term, the logarithm of the 1-octanol/water partition coefficient (log K_{OW}), for each chemical were secured as either a measured or computer-estimated value from the ClogP for Windows software (BIOBYTE Corp., Claremont, CA, USA); the measured value was used in preference to a calculated value.

For determination of molecular orbital descriptors each toxicant was built and energy minimized in the NEMESIS for PC molecular modelling software (Oxford Molecular Limited, Oxford, UK). Compounds were saved as .pdb (protein database) files and converted into MOPAC internal files via the BABEL shareware file conversion program. Molecular orbital quantum chemical calculations were performed using the MOPAC6 program (Stewart 1990). Each molecule was geometry optimized using the AM1 Hamiltonian in MOPAC6. The following keywords were employed: AM1 ENPART GEO-OK NOXYZ NOINTER PRECISE. The energy of the lowest unoccupied molecular orbital (E_{LUMO}) was obtained for each compound.

10.2.4   DEVELOPMENT OF RESPONSE-SURFACE

Quantitative structure–activity relationships (QSARs) were developed using the MINITAB (version 10.1) statistical package. For QSAR development, log (IGC_{50}^{-1}) acted as the dependent variable. The quantifiers of hydrophobicity and electrophilicity acted as the independent variables. Model adequacy was quantified with the r^2 value (coefficient of determination). The s-value (root of the mean square for error), the F-value (Fisher statistic) for the whole regression equation, and Q^2 (leave-one-out cross-validation coefficient of determination) were also noted. The Q^2 value was calculated using a macro in the MINITAB (version 10.1) statistical package (Andrew Worth, personal communication). Outliers were established with reference to their residuals and Studentized deleted residuals.

10.3   MODELLING TOXICITY TO *TETRAHYMENA PYRIFORMIS*

The toxicities and calculated physico-chemical parameters of the aliphatic carbonyl-containing compounds are listed in Table 10.1.

The relationship between toxicity and hydrophobicity is shown in Figure 10.1. As would be expected, the ketones and mono-esters (shown as solid circles in Figure 10.1), which are known to act as non-polar narcotics, form a baseline, or minimum, toxic effect. The other carbonyl-containing compounds have toxicity in excess of this baseline. This excess toxicity can be assumed to be as a result of covalent interactions between the toxicant and biological macromolecules. Consideration of the chemistry of these molecules (see Table 10.2) would suggest that there are a number of different potential mechanisms of toxicity.

Attempts to model toxicity will therefore require both parameters to account for penetration and for interaction. The QSAR based on these two parameters, for all compounds is shown below.

$$\log (IGC_{50}^{-1}) = 0.560 \log K_{OW} - 1.03\ E_{LUMO} - 0.403$$
$$n = 140,\ r^2 = 0.753,\ s = 0.500,\ F = 209,\ Q^2 = 0.740 \tag{4}$$

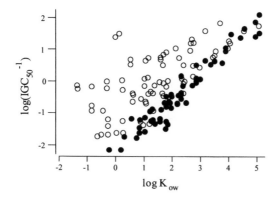

Figure 10.1 Plot of toxicity to *Tetrahymena pyriformis* (log (IGC_{50}^{-1})) against hydrophobicity (log K_{OW}). Solid circles indicate ketones and mono-esters, which are considered to act as non-polar narcotics.

While the t-values on each variable (18.3 and -12.9 respectively), and the F-statistic are significant, this QSAR exhibits only modest goodness-of-fit and predictivity.

Examination of the residuals from the predicted fits suggests that there are five outliers to equation (4). These compounds (acrolein; 3-buten-2-one; β-butyrolactone; β-propiolactone; trans-ethyl crotonate) had toxicity much greater than predicted by equation (4). Such compounds are characterized as being small and intrinsically reactive.

Removal of outliers from any model, simply to improve statistical fit, should be performed only cautiously. While outliers (which may be defined by a number of statistical criteria including the standard residual) may be removed from QSAR, this should not be attempted *ad infinitum*. Despite this, it is often considered that analysis of the outliers from an investigation can provide additional information (Lipnick 1991). What is favoured is a removal of outliers following a pragmatic approach. In this study there are sound biological and chemical reasons for omitting the five compounds listed. Removal of these outliers results in the following highly significant, and predictive, QSAR:

$$\log{(IGC_{50}^{-1})} = 0.620 \log K_{OW} - 1.04\ E_{LUMO} - 0.55$$
$$n = 135,\ r^2 = 0.853,\ s = 0.385,\ F = 382,\ Q^2 = 0.846 \tag{5}$$

Both variables are significant in this equation with t-values of 25.2 and -16.4 respectively.

There were also a number of outliers to equation (5). These compounds included 2-butenal, which had toxicity greater than predicted by equation (5). As with outliers to equation (4), 2-butenal is characterized as being small, and intrinsically reactive. Other compounds (e.g. 2,4-dimethyl-2,6-heptadienal) also were found to be outliers; with these molecules the observed toxicity less than predicted by equation (5). It is postulated that in these cases steric hindrance inhibits reactivity. For example, the reactive centre of 2,4-dimethyl-2,6-heptadienal (an α,β-unsaturated aldehyde) is clearly sterically hindered by the methyl group in the 2-position.

Comparisons of the slopes on both variables, as well as the intercept, between equations (3), for aromatic compounds, and (5), for aliphatic compounds, demonstrate a significant difference between these two QSARs. This may suggest that while the same generic mechanism of toxic action occurs for these compounds, equations (3) and (5) describe separate 'domains' within the response-surface for *Tetrahymena* toxicity. A further complication is that E_{LUMO} values may be intrinsically incomparable for aliphatic and aromatic compounds, with E_{LUMO} values being much lower for the π-bond rich aromatic compounds.

Toxicity is presented by the chemicals used in this study as a continuum. Narcotic chemicals (e.g. ketones) showing no evidence of electrophilic reactive toxicity are distributed along the toxicity-log K_{OW} face. Chemicals of moderate electrophilicity (e.g. α,β-unsaturated ketones) are distributed in the central area of the plane, whereas, the most electrophilic chemicals (i.e. α-diones) are observed in the upper part of the plane characterized with the lowest E_{LUMO} values.

The response-surface approach described in this study is a specific tool to predict toxicity that defines *a priori* the descriptors to be hydrophobic and molecular orbital parameters. The strength of the surface-response approach for the prediction of toxicity is that it circumvents the direct identification of a mechanism of toxic action, or assignment of a chemical to a particular chemical class. This is required in some 'popular' QSAR approaches to predicting toxicity, most notably the ECOSAR expert system utilized by the United States Environmental Protection Agency (Clements and Nabholz 1994). Current efforts envisage the development of toxicity prediction using a response-surface approach in conjunction with a rule-based expert system. Such a rule-based system could identify, and highlight, those compounds considered to be outliers from the response-surfaces.

10.4 CONCLUSIONS

To conclude, a two-parameter QSAR, or response-surface, has been developed to describe the toxicity to *Tetrahymena* of aliphatic compounds, which are known to be acting by a variety of mechanisms of action. A small number of well-defined outliers were observed. Therefore, this method does provide a means of forecasting the toxic effects of chemicals from a consideration of chemical structure alone. QSARs can play an important role in predicting the toxicity of chemicals to environmentally important species and therefore should be incorporated into the forecasting of the potential environmental impact of chemicals released into the environment.

REFERENCES

Bradbury SP and Lipnick RL (1990) Introduction: Structural properties for determining mechanisms of toxic action. *Environmental Health Perspectives*, **87**, 181–182.

Clements RG and Nabholz JV (1994) *ECOSAR: A Computer Program and User's Guide for Estimating the Ecotoxicity of Industrial Chemicals Based on Structure Activity Relationships*. EPA-748-R-93-002, United States Environmental Protection Agency, Washington DC, USA.

Cohen SD, Pumford NR, Khairallah EA, Boekelheide K, Pohl LR, Amouzadeh HR and Hinson JA (1997) Selective protein covalent binding and target organ toxicity. *Toxicology and Applied Pharmacology*, **143**, 1–12.

Cronin MTD (1998) Computer-aided prediction of drug toxicity in high throughput screening. *Pharmacy and Pharmacology Communications*, **4**, 157–163.

Cronin MTD and Dearden JC (1995) Review: QSAR in toxicology. 1. Prediction of aquatic toxicity. *Quantitative Structure-Activity Relationships*, **14**, 1–7.

Cronin MTD, Gregory BW and Schultz TW (1998) Quantitative structure–activity analyses of nitrobenzene toxicity to *Tetrahymena pyriformis*. *Chemical Research in Toxicology*, 11, 902–908.

Cronin MTD and Schultz TW (1998) Structure-toxicity relationships for three mechanism of toxic action to *Vibrio fischeri*. *Ecotoxicology and Environmental Safety*, **39**, 65–69.

Goldschmidt MC and Lockhart BM (1971) Simplified rapid procedure for determination of agmatine and other guanidino-compounds. *Analytical Chemistry*, **43**, 1475–1479.

Hermens J (1991) QSAR in environmental sciences and drug design. *Science of the Total Environment*, **109/110**, 1–7.

Hermens JLM (1990) Electrophiles and acute toxicity to fish. *Environmental Health Perspectives*, **87**, 219–225.

Jaworska JS, Hunter RS and Schultz TW (1995) Quantitative structure-toxicity relationships and volume fraction analyses for selected esters. *Archives of Environmental Contamination and Toxicology*, **29**, 86–93.

Karabunarliev S, Mekenyan OG, Karcher W, Russom CL and Bradbury SP (1996) Quantum-chemical descriptors for estimating the acute toxicity of electrophiles to the fathead minnow (*Pimephales promelas*): An analysis based on molecular mechanisms. *Quantitative Structure–Activity Relationships*, **15**, 302–310.

Karcher W and Devillers J (eds) (1990) *Practical Applications of Quantitative Structure-Activity Relationships (QSAR) in Environmental Chemistry and Toxicology*, Kluwer, Dordrecht, Netherlands.

Könemann H (1981) Quantitative structure–activity relationships in fish toxicity studies. Part 1: Relationship for 50 industrial pollutants. *Toxicology*, **19**, 209–221.

Lipnick RL (1991) Outliers: their origin and use in the classification of molecular mechanisms of toxicity. *Science of the Total Environment*, **109/110**, 131–153.

Martin YC (1978) *Quantitative Drug Design*. Marcel Dekker, New York, USA.

McFarland JW (1970) On the parabolic relationship between drug potency and hydrophobicity. *Journal of Medicinal Chemistry*, **13**, 1092–1196.

Mekenyan OG and Veith GD (1994) The electronic factor in QSAR: MO-parameters, competing interactions, reactivity, and toxicity. *SAR and QSAR in Environmental Research*, **2**, 129–143.

Schultz TW (1997) TETRATOX: *Tetrahymena pyriformis* population growth impairment endpoint: A surrogate for fish lethality. *Toxicology Methods*, **7**, 289–309.

Schultz TW and Cronin MTD (1999) Response-surface analyses for toxicity to *Tetrahymena pyriformis*: reactive carbonyl-containing aliphatic chemicals. *Journal of Chemical Information and Computer Sciences*, **39**, 304–309.

Schultz TW and Deweese AD (1999) Structure–toxicity relationships for selected lactones to *Tetrahymena pyriformis*. *Bulletin of Environmental Contamination and Toxicology*, **62**, 463–468.

Schultz TW, Sinks GD and Cronin MTD (1997) Identification of mechanisms of toxic action of phenols to *Tetrahymena pyriformis* from molecular descriptors. In *Quantitative Structure–Activity Relationships in Environmental Sciences— VII*, Chen F and Schuurmann G (eds), SETAC, Pensacola, FL, pp. 329–342.

Stewart JJP (1990) *MOPAC Manual* (6th edn), Frank J. Seiler Research Laboratory, US Air Force Academy, Colorado Springs, CO, USA.

van Wezel AP and Opperhuizen A (1995) Narcosis due to environmental pollutants in aquatic organisms: Residue-based toxicity, mechanisms, and membrane burdens. *Critical Reviews in Toxicology*, **25**, 255–279.

Veith GD, Call DJ and Brooke LT (1983) Structure–toxicity relationships for the fathead minnow, *Pimephelas promelas*: Narcotic industrial chemicals. *Canadian Journal of Fishery and Aquatic Science*, **40**, 743–748.

Veith GD and Mekenyan OG (1993) A QSAR approach for estimating the aquatic toxicity of soft electrophiles (QSAR for soft electrophiles). *Quantitative Structure–Activity Relationships*, **12**, 349–356.

Application of a Transgenic Stress-inducible Nematode to Soil Biomonitoring

ROWENA S. POWER AND DAVID I. DE POMERAI

School of Biological Sciences, University of Nottingham, Nottingham, UK

11.1 INTRODUCTION

The measurement of the effect of toxic chemicals on organisms is most commonly studied using indicators such as lethality, growth and reproduction. Most toxicant effects begin as interactions with biomolecules, however, so the study of effects at the biochemical level may indicate effects at the higher levels of organization, acting as 'biomarkers'. The major advantage of using biochemical and cellular effects as a biomarker is that they tend to be sensitive, less variable, highly conserved and easier to measure than whole organism effects (Sanders 1990). Biomarkers that are involved in protecting the cell and defending it from environmental insults represent a response at the biochemical level that can hopefully be related to the health of the organism and subsequently adverse effects on the whole population. Thus the measurement of a biomarker has the capacity to allow the recognition of degree of toxic effect being exercised by a contaminant in the environment. Such a biomarker can therefore contribute to forecasting the environmental potential of any change in the availability of the contaminant, not just in terms of the organism concerned but also up the scale to processes at the level of population and community.

The heat shock or stress response is common to all living organisms and is characterized by the induction of a unique set of polypeptides called heat shock proteins (HSPs). Heat shock proteins are induced as a general response to protein damaging events such as heat (hence the name), metals and other toxicants, microwave radiation and ultraviolet light. The binding of heat shock factor (HSF) to the heat shock element (HSE) switches on the transcription of stress-inducible *hsp* genes, but the mechanism by which the heat or other stress activates the HSF is obscure. It is known that HSF is activated in the presence of damaged proteins. Indeed, the function of the stress protein produced is to repair or inactivate the damaged proteins so that they do not cause further cellular

Forecasting the Environmental Fate and Effects of Chemicals. Edited by Philip S. Rainbow, Steve P. Hopkin and Mark Crane.
© 2001 John Wiley & Sons Ltd

damage (Sanders 1993). They provide protection from environmentally induced damage and confer tolerance to the stressor at sublethal levels.

HSPs from a number of organisms have been evaluated as biomarkers of pollution (reviewed by Sanders 1993, de Pomerai 1996). There has been considerable interest in stress responses in the soil nematode worm *Caenorhabditis elegans*. *C. elegans* lends itself well to such applications as it is one of the most well-studied animals at the genetic, physiological, molecular and developmental levels. Guven and de Pomerai (1995) looked at HSP70 responses to metals in *C. elegans*. Separation of total worm protein on two dimensions (pI and MW) allowed individual proteins to be detected, and when probed with anti-HSP70 antibody distinctive induction patterns were seen. This approach is time consuming and costly however, and several gels of the same sample have to be run in order to recognize the distinctive pattern. The development of transgenic *C. elegans* strains with reporter genes fused to the heat shock promoter regions has simplified studies of stress-response induction (see Fire 1986, Stringham *et al.* 1992) and strains carrying a stress-inducible β-galactosidase reporter have been assessed for toxicity testing (Stringham and Candido 1994, Guven *et al.* 1994). In strain PC72 a β-galactosidase gene is fused in-frame within the *hsp*16-1 gene. Thus when HSP16 is induced, β-galactosidase is also produced. This can be quantitatively measured by a simple enzyme assay (Dennis *et al.* 1997) or located within the worm using histochemical stains (Guven *et al.* 1994, Mutwakil *et al.* 1997). To date the PC72 strain has been used for evaluating the response to particular chemicals (Stringham and Candido 1994, Guven *et al.* 1994, Jones *et al.* 1996) and mixtures of metals and surfactants (Dennis *et al.* 1997) in simple water tests and has been applied as an ecotoxicological monitor of river samples (Mutwakil *et al.* 1997). More novel applications include its use for examining the stress induced by the mammalian immune response (Nowell *et al.* 1997), and as an indicator of microwave radiation-induced stress (Daniells *et al.* 1998).

Recent work (Power *et al.* 1998) examined the potential application of PC72 *C. elegans* to soil biomonitoring. *C. elegans* has been used for studies of soil and sediment toxicology, using LC_{50} (Donkin and Dusenbery 1993) and growth and reproduction (Traunsperger *et al.* 1997) end points. PC72 *C. elegans* represents an organism that could be applied to rapid ecotoxicological assessment, in a similar manner to the bacterial Microtox™ test, but as a multicellular organism and next stage of the food chain (it is bacteriverous), it could complement and extend the information current methods provide.

The induction of the stress response to soluble and insoluble metal additions to soil (Power *et al.* 1998) and metal mixtures (Power and de Pomerai 1999) was carried out in one soil type. This work established that exposure of the worms induced a response that was proportional to the response in worms exposed in water with metals added at the soluble concentration found in the soil water (Power and de Pomerai 1999). However, the bioavailability of a metal to the organism is not limited to the soluble portion: it was also found that insoluble salts added to the soil also induced a response and that food bacteria could play an important role in passing metals on to the worm (Power *et al.* 1998). The bioavailability of metals is subject to many factors such as pH, organic matter, clay content and the presence of other metals; therefore, the results are limited by only having been done in a sandy loam, and it would be appropriate to extend the study using other soil types.

Colleagues at the University of Nottingham have set up microcosms composed of soils representing a range of soil types, pH and organic matter to monitor long term the bioavailability of added metal mixtures representing current UK agricultural sludge limits for soils. This has provided a useful source of material to further determine the suitability of this test in the study of soil pollution by exposing worms to the same metal concentrations, but in different soil types.

This chapter describes modifications to the test for soil toxicity testing and briefly describes the initial results of investigations using these soils. This leads on to a discussion of the relevance of stress responses and of the suitability of the test for application to soil testing with the ultimate aim of establishing a biomarker of value to soil ecotoxicologists.

11.2 EXPERIMENTS WITH *CAENORHABDITIS ELEGANS*

11.2.1 THE WORMS

The PC72 transgenic strain of *C. elegans* (carrying a *C. elegans hsp16* promoter/*E. coli lacZ* fusion gene) and the *C. elegans hsp*16-2 polyclonal anti-rabbit antibody were kindly provided by Professor E.P.M Candido (Department of Biochemistry and Molecular Biology, University of British Columbia, Vancouver, Canada) (Stringham and Candido, 1993, 1994). The *lac*-deleted *E. coli* strain P90C was a generous gift from Dr A. Chisholm (MRC Molecular Biology Laboratory, Cambridge, and UK).

PC72 *C. elegans* were cultured on worm growth medium (NGM) agar with a lawn of *E. coli* P90C (Guven *et al.* 1994). Old plates with mixed populations were divided up and added to fresh agar for establishment of new plates. After three to four days growth at 15 °C, the worms were washed off with K medium (53 mM NaCl, 32 Mm KCl; Williams and Dusenbery 1990), filtered through a 5 μm Wilson Sieve (University of Nottingham, Nottingham, UK) to remove L1 and L2 larvae and collect young adults. The adults were washed off the filter and collected for addition to the soil samples. Normally about 1×10^4 worms were used in each 1 g soil sample or equivalent.

11.2.2 SOIL SAMPLES

The sandy loam used for standardizing the test was Lufa 2.2, a well-characterized commercially available soil from Landwirtschaftliche Untersuchs und Forschungsanstalt (Speyer, Germany). This soil was air dried and stored at room temperature and prior to worm exposure wetted to 50% moisture-holding capacity (MHC). The test soils were kindly provided by Dr Andy Tye, University

Table 11.1 Metal additions to soil samples (Tye and Young, personal communication)

Element	UK limit[a] (mg/kg)	Salt used for addition
Zn	300	$Zn(NO_3)_2.6H_2O$
Cu	135	$Cu(NO_3)_2.3H_2O$
Ni	75	$Ni(NO_3)_2.6H_2O$
Cd	3	$Cd(NO_3)_2.4H_2O$
Pb	300	$Pb(NO_3)_2$
As	50	$Na_2HAsO_4.7H_2O$

[a]Sludge (use in Agriculture) Regulations limits 1989 (EC directive 86/278) for soil in pH 6.0–7.0 band.

of Nottingham. Soils were collected to a depth of 23 cm in June 1997, sieved to 4 mm and maintained at 50% MHC. Metals were applied at rates equivalent to UK agricultural sludge limits for soils in a 6–7 pH range (Table 11.1) and incubated at 16 °C. Control soils with no added metals were also set up and maintained in these conditions. Samples were taken for this study after eight months' incubation and stored at 4 °C until used, but the microcosms continued to be incubated for long-term analysis of metal availability by Dr Andy Tye and Dr Scott Young, University of Nottingham.

11.2.3 WORM EXPOSURE

Soil samples were prepared by adding *E. coli* P90C at approximately 10^8 CFU g^{-1} soil (overnight culture grown up in LB at 37 °C, then stored at 4 °C until used, pelleted by centrifugation at 1500 **g** for 10 min, then resupended in 0.2 M phosphate buffer (pH 7) for addition to soil). Lufa 2.2 soil, used for standardizing the methods also received additions of $CdCl_2$, $HgCl_2$, $CuCl_2$ or $ZnCl_2$ at this point. Initial experiments were performed with 1 g of soil in 15 ml centrifuge tubes, but latterly in 24-well tissue culture plates. The prepared soil was incubated for 24 h at 25 °C to equilibrate before adding the worms collected as above. The worms were exposed for a 24 h incubation period at 25 °C, then killed by freezing in the test samples. The soil was maintained at 50% MHC throughout the equilibration period and exposure of the worms.

11.2.4 RECOVERY OF WORMS FROM SOIL

The worms were recovered using a silica suspension (Ludox-HS40, Aldrich) as described by Donkin and Dusenbery (1993), but with some modifications. The soil was moistened with 0.2 M phosphate buffer to maintain the pH at 7 and free the worms. A 1:1 (v/v) dilution of Ludox-HS40 (pH corrected to 7.0) was mixed well with the soil and centrifuged at 700 **g** for 1 min. The upper liquid layer, with the floating worms, was drawn off and filtered through a 5 μm mesh allowing the Ludox to drain off and the worms to be washed and collected.

11.2.5 MEASUREMENT OF β-GALACTOSIDASE EXPRESSION

The β-galactosidase produced by the induction of the $hsp16/lacZ$ construct was measured by analysis of its activity. 4-methylumbelliferone (4-MU) is the fluorescent product β-galactosidase digestion of 4-methylumbelliferyl-β-D-galactopyranoside and can be quantitatively measured on a DyNA Quant™ 200 Fluorometer (Hoefer Pharmacia Biotech) following the method of Dennis et al. (1997). This method was also modified for use with the Millipore Multiscreent™ system. The worms recovered from each sample were placed in the well of a 96 well Multiscreen plate with 5 μm, solvent resistant mesh. Using the Millipore vacuum manifold, any liquid carried over with the worms was drawn off through the mesh, leaving the worms in the wells. The worms were then permeabilized using acetone, which was also drawn off. 50 μl of buffer (25 mM Tris-HCl, pH 7.5; 125 mM NaCl; 2 mM $MgCl_2$; 10 mM DTT), containing 0.1 mg ml^{-1} of the substrate was added to each well and the plate incubated for 30 min at 37 °C. The reaction was stopped by adding 12.5 μl of 25% (w/v) trichloroacetic acid (TCA) into each well, then drawing off the MU-containing supernatants into a non-fluorescent black 96-well plate containing glycine-carbonate buffer. The 4-MU product was quantified on a Perkin Elmer HTS7000 Fluorescent plate reader, each run being standardized with a stock solution of 50 nM 4-MU. After enzyme activity measurements, the worms on the Multiscreen plate were washed with 5% (w/v) TCA and protein content determined using a modified Lowry protein determination (Lowry et al. 1951) involving solublization of the worms in 10 M NaOH for 1 h prior to the addition of 2% (w/v) $NaHCO_3$ in place of the NaOH/Na_2CO_3 Lowry A solution (Dennis et al. 1997). β-galactosidase activity was expressed as pmoles 4-MU h^{-1} μg protein^{-1}.

11.2.6 TWO-DIMENSIONAL GEL ELECTROPHORESIS AND WESTERN BLOTTING

Worms from 5–10 plates of worms were exposed in the test samples and incubated and frozen as before. The worms were homogenized thoroughly in 100–300 μl of homogenization buffer (1 mM dithiothreitol (DTT), 0.1% SDS, made up in TBS (154 mM NaCl, 10 mM Tris-HCl, pH 7.5)), then protein content determined by the method of Lowry et al. (1951). 1–50 μg of protein was mixed in at least a 1:4 ratio with sample solution buffer (9 M urea, 65 mM DTT, 0.5% Triton X-100, 2% Pharmalyte 3–10™ and a few grains of Bromophenol blue (BPB)). Samples were run using the Immobiline™ Drystrip Kit in the first dimension and ExcelGel™ precast SDS-gradient gels in the second dimension (both from Pharmacia Biotech). The first dimension, isoelectric focusing, was run on 11 cm, pH 3–10 drystrips, and the second dimension on a 8–18% gradient gel following manufacturer's instructions. After running, the gel backing was removed and the proteins transferred on to ECL nitrocellulose membrane (Amersham) (Guven and de Pomerai 1995). After blotting, the total protein was detected using Protogold stain (British Biocell International), then blocked

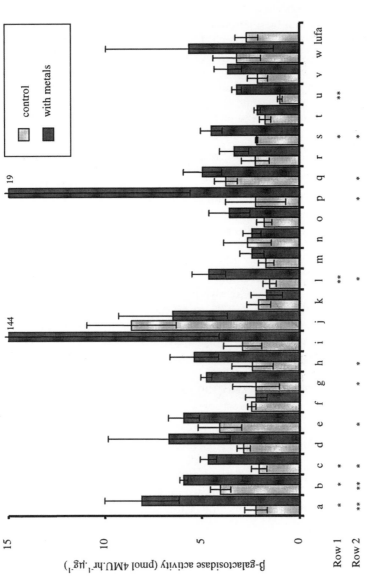

Figure 11.1 β-galactosidase activity in PC72. *C. elegans* exposed for 24 h at 25 °C to a variety of soils with and without metal conditions. Metal additions are described in Table 11.1 and soil characteristics in Table 11.2. Column heights represent the mean activity of three samples, and the bars the standard error of the mean. Responses in worms exposed to metal treated soil that are significantly different from those exposed to control treatments of the same soils are shown in Row 1. Row 2 shows responses in worms exposed to metal treated soil that are significantly different from those in worms exposed to Lufa 2.2 soil with no metal additions (* = $p < 0.05$, ** = $p < 0.01$).

Table 11.2 Properties of the soils studied

Soil	Soil type	Land use	pH	% organic matter
a	Brown sand	Arable	5.85	2.97
b	Stagnogley	Arable	5.81	2.97
c	Gleyic brown earth	Arable	6.04	2.72
d	Ranker/brown podzolic	Heathland	3.29	11.68
e	Typical brown earth	Deciduous woodland	3.09	13.87
f	Typical brown earth	Arable	6.16	8.81
g	Pelo-vertic alluvial gley	Pasture	5.55	
h	Pelo stagnogley	Arable	6.81	8.24
i	Typical stagnogley	Deciduous woodland	2.88	6.67
j	Argillic pelosol	Arable cereals	6.80	
k	Cambic stagnogley	Grass	5.93	11.04
l	Typical brown earth	Grass	5.97	8.45
m	Silty loam	Arable	5.16	
n	Loamy sand	Arable	5.48	0.86
o	Humic rendzina	Shrub/grass	6.81	
p	Pelo stagnogley	Arable	7.06	9.46
q	Calcareous pelosol	Grass	6.80	4.70
r	Calcareous pelosol	Arable	7.08	7.03
s	Silty clay loam	Arable	5.93	
t	Sandy loam	Arable	6.23	
u	Clay loam	Arable	4.85	
v	Sandy loam	Arable	5.86	
w	Silty clay loam	Arable	6.89	

*Information provided by Andy Tye and Scott Young.

overnight at 4 °C in TBS/0.1% Tween 20 (TBS-T) with 5% bovine serum albumin (BSA). The primary antibody treatment was a 1:2500 dilution of the *hsp*16 antibody in TBS-T with 5%BSA for 1 h, and secondary with a 1:5000 dilution of horseradish peroxidase (HRP) anti-rabbit IgG for 1 h. The antibody complexes were determined using the ECL chemiluminescent detection system.

11.3 REPORTER RESPONSE

11.3.1 SOIL TESTING

In assessing the reporter response to exposure in different soil types, 23 soils were tested, in all cases comparing control with metal-containing versions. The results in Figure 11.1 show the β-galactosidase activity induced after exposure to the soils described in Table 11.2. Worms recovered from soil (i) with metal showed a large response, but with great variation, making it not significantly different from the worms exposed in soil (i) without metal additions. Figure 11.1 also shows whether the responses in worms exposed to metal-treated soils were significantly different from the responses to those exposed to the control soils from the same sites (row 1) or to Lufa 2.2 standard soil (row 2). Among the control soils, only responses in worms exposed to soil (j) differed significantly from those exposed to Lufa 2.2 control soil ($p < 0.05$).

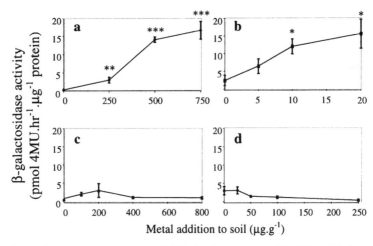

Figure 11.2 β-galactosidase activity in PC72. *C. elegans* exposed for 24 h at 25 °C in Lufa 2:2 soil with metal conditions. (a) Cd (as $CdCl_2$); (b) Hg (as $HgCl_2$); (c) Cu (as $CuCl_2$); (d) Zn (as $ZnCl_2$). Points represent the mean activity of three samples, and the bars the standard error of the mean.

Figure 11.2 shows the response of worms to overnight exposure in Lufa 2.2 soil with single metal additions. Additions of cadmium above $250\,\mu g\,g^{-1}$ resulted in a significantly different β-galactosidase activity ($p < 0.01$). This response levelled off at $1000\,\mu g\,g^{-1}$, after which it dropped. Mercury additions induced a response at lower levels than are required with cadmium and a significant response was seen at $10\,\mu g\,g^{-1}$ ($p < 0.05$). Worms exposed to copper and zinc showed no significant responses over the ranges shown here. The worms were observed to be dead at the end of the incubation period in higher levels of copper.

Figure 11.3 confirms the induction of HSP16 by exposure in soil with cadmium additions. The Western blot shows the induction of a set of spots in the 16 KDa region in the blots of protein from worms that had been exposed to Lufa soil with cadmium that were not present in the blot of protein from worms exposed to soil without metal additions.

11.4 DISCUSSION AND CONCLUSIONS

11.4.1 EXPOSURE TO METAL MIXTURES

The worms exposed to soils with metal mixtures did not consistently show a greater stress response than the worms exposed in the control soils, but this should perhaps not be too surprising, since the concentration at which each metal was applied is the maximum for agricultural soils, implemented as a legislative safeguard. An example of one of the exceptions is the response of the worms exposed to the soil (a) with metal additions, which showed a significant difference from the worms exposed in the control soil. Soil (a) did not possess any

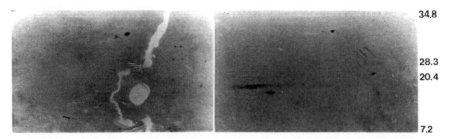

Figure 11.3 Two-dimensional western blots of stress-induced HSP16 proteins after exposure of the PC72. *C. elegans* to Lufa 2.2 soil (a) and Lufa 2.2 soil with 250 μg g^{-1} cadmium additions (b). Protein samples were run on two dimensions, as described in the text, then transferred to nitrocellulose membrane and probed with rabbit anti-*C. elegans* HSP16 antibody, followed by anti-rabbit horseradish peroxidase linked secondary antibody and chemiluminescent detection. Figures mark the position of markers (kDa).

distinctive features; the pH was 6.57, in the middle of the range tested. It had a low organic matter content and is sandy, which could lead to high metal concentrations in the soil water solution, but the measured metal concentrations were not among the highest (Tye and Young personal communication). This result could either cast doubt on the usefulness of this test, or could highlight an unexpected problem where one or more metals is significantly more bioavailable to the worms than measured concentrations in the soil water would suggest. Tests with other biomarker systems could confirm whether or not this soil is indeed significantly more toxic than might be expected.

Soil (i) with metal additions was a very interesting sample that might have been predicted to induce a strong response in the worms. This soil was collected from mine spoil, with a high iron pyrites content, oxidizing to sulphuric acid and giving a very low pH of 3.36. A very high response was seen in one of the samples and in repeated tests the response was sometimes significant, but the data set used here was more representative of the results found from repeated experiments. This could reflect problems with the storage of the samples (affecting the pH through variable oxidation rates) or the high levels of metal and/or the low pH could be killing the worms before a stress response could be seen. Worms were not routinely examined for viability at the end of each assay, but were frozen after exposure to prevent heat shock of the live worms while recovering them.

Measurements were difficult to obtain from worms exposed to soils with a high clay or organic matter content. When recovering the worms after exposure, the organic matter floated and made it difficult to obtain a clean sample of worms. Recovery from clay soils was also difficult, and the responses tended to be low. This may be due to difficult handling properties, leading to problems such as insufficient mixing, or may reflect the protective property of clay soils.

However, looking for differences between worms exposed in the same soils with different treatments may not be a good assessment of the test. All the control soils varied in the response they induced in the worms, so there is no 'baseline' of uncontaminated soils to work from, and there are no data for the metal concentrations in the soil before metal additions to support a correlation with the responses shown. A useful observation from the results in control soil (no metal additions) is that only one shows a significantly different response from the response in Lufa soil. This suggests that Lufa soil may be useful as a basis for comparison.

There is a relationship between cadmium concentration in the soil water solution and the stress response in the worms exposed to the metal treated soils described in Table 11.2. These data are not presented here, but will be presented in a future publication. However, it can be noted that in general the rank order of cadmium concentration in soil water is the reverse of the order of pH in the soils studied here (see Table 11.2), i.e. cadmium concentration increases with soil acidity, as expected.

11.4.2 RESPONSE TO SINGLE METAL ADDITIONS

A high concentration of cadmium was required to elicit a response in the worms in soil tests, yet in aquatic tests, cadmium is considered to be a strong inducer of the stress response at $8-16 \mu g \, ml^{-1}$ (Guven et al. 1994). Since Lufa 2.2 soil has a relatively high organic matter content (2.3% organic carbon), cadmium could be more strongly adsorbed, and higher quantities may be required to allow sufficient cadmium to come into solution. The adsorption behaviour of metals is determined by the pK (equilibrium constant) values of the reaction $M^{2+} + H_2O = MOH^+ + H^+$. The basic order of motility is

$$Cd(pK = 10.1) < Ni(pK = 9.9) < Co(pK = 9.7) < Zn(pK = 9.0)$$
$$\ll Cu(pK = 7.7) < Pb(pK = 7.7) < Hg(pK = 3.4) \text{ (Alloway 1990)}$$

While cadmium is relatively low in the pK order, Elliot et al. (1986) demonstrated that organic matter restricts cadmium mobility more than would be anticipated. AAS analysis showed that when the total soil concentration of cadmium is $100 \mu g \, g^{-1}$, the soil water concentration is around $5 \mu g \, ml^{-1}$ (Power and de Pomerai 1999). This result highlights the problems of studying such responses in only one soil type. Concerns that the test is too insensitive due to the high concentrations required to produce a clear response can be allayed to some extent by the consideration that pollution with a single metal is unlikely to occur, which is one reason why mixed inputs were studied. Lower concentrations of mercury are required to induce the stress response, but this is also found in water tests. Copper concentrations of at least $10 \mu g \, ml^{-1}$ are required in the water tests to produce a significant response (unpublished observations), and no stress responses have been detected to any level of copper added to Lufa 2.2 soil. Like copper, zinc on its own can only elicit a stress response in worms when administered at much higher aqueous concentrations than cadmium ($>50 \mu g \, ml^{-1}$), so it is not surprising that no induction is seen in soil.

11.4.3 INTERACTIONS OF METALS

It may not be suitable to just examine the response of the worm to one metal or a number of them together; possible antagonistic or synergistic effects of the metals may occur and all possible interactions should be considered. Combinations of cadmium and copper in Lufa 2.2 soil induced a stress response at lower concentrations of cadmium than would be expected (Power and de Pomerai 1999). This was thought to be due to preferential adsorption of the copper to the soil components, allowing more cadmium to remain in solution. Posthuma *et al.* (1997) found that copper additions to their artisol had a similar effect on the zinc concentration in soil water. It was found though that zinc inhibited the stress response of the PC72 *C. elegans* to cadmium, but appeared to interact similarly with the soil constituents resulting in an increase of both metals in solution with increased additions to the soil (Power and de Pomerai 1999). This implies that there are important uptake and intracellular factors also involved in the initiation of the stress response due to these metals. Guven *et al.* (1995) found that zinc was among the metals competing with cadmium for entry into the *C. elegans* through calcium channels. These effects cannot be generalized however; Doelman *et al.* (1984) found low concentrations of lead and cadmium together inhibited reproduction of their test worms, *Mesorhabditis monohystera* and *Aphelenchus avenae*, less than the additive effect expected after five days, yet at higher levels of the metals a synergistic effect was measured.

11.4.4 THE BIOAVAILABLE PORTION

Chemical analysis provides abundant information on actual metal concentrations in the soil water solution. There are many sensitive and accurate chemical analyses for metal ions, but the actual bioavailability to a particular organism cannot be predicted for all stages of interaction. Competition between the metals for adsorption sites in the soil (as discussed) is only one interaction that could affect the eventual response of an organism. A concept originally proposed by Calamari and Alabaster (1980) for examining the effect of mixtures of toxicants in aquatic environments has been applied by van Gestel and Hensbergen (1997) to the soil environment. There are three potential interactions: the interaction with soil constituents as described previously, interactions affecting uptake from the soil solution and therefore the amount entering the organism and interactions at the target site within the organism. At any of these stages interactions between metals may occur, an important consideration since metal-polluted soils often contain a mixture of pollutants.

The microbial population of the soil also has an important role in balancing immobilization and mobilization of metals (Chanmugathas and Bollag 1987). Over an eight-week period a significant amount of cadmium was released into soil water when bacteria were present, whereas none was found in soil water when the soil was sterile. In our study the soils are only incubated for a total of 48 h, and yet the bacteria still had a major influence on the results (Power *et al.* 1998) and it was also found that bacteria grown up in media with cadmium additions passed on the metal and increased the stress response in worms. Doelman *et al.* (1984) found high metal restricted fungal growth in their fungi to worm food chain, resulting in a lack of food for the nematode. If measuring death or reproduction end points of toxicology, this would be important; in our work the stress response is only a little affected by food availability. The only way this can affect the test is if the worms die and hence no response is induced. There remains a question as to whether food bacteria should be included in this test as they could pass on the metals through the worm's digestion or alternatively remove metals from soil solution, thereby increasing or decreasing metal bioavailability.

11.4.5   APPLICATION OF THE TEST AS A RAPID ASSAY

The description of methods outlines some of the developments we are making towards a rapid assay using the PC72 *C. elegans*. The use of multiwell plates and plate readers has accelerated our operations considerably. The major obstacle in applying the test to soil is the time taken to recover the worms from soil. The use of extracted soil water for testing simplifies the methodology considerably, which could be useful for future development of a rapid test. However, there is a slight difference in response, which could be a potential problem in the replacement of soil tests with tests in soil water. In a long-term experiment the use of soil water would not be realistic, since exchange between soil water and adsorption sites, which would occur in soil, could not occur if the soil water were removed. An interesting finding from previous work is that the omission of *E. coli* in the soil exposures reduces the response of the worms in much the same way as the exposures in soil water (Power *et al.* 1998).

A major disadvantage of work with transgenic worms is that they cannot be released in the field. Test kits could be designed to contain them for water testing, but for soil biomonitoring, samples would have to be taken to a laboratory.

Only one effect on the worm is examined in the work discussed, namely that of the stress response, even more specifically the induction of *hsp*-16 whose promoter regulates the transgene. It is apparent from our results that cadmium induces the response at lower input levels than either copper or zinc. Analysis of other biomarker responses will permit consideration of a number of other effects of the metals, at various sites and on various target functions. The effect of cadmium additions on *C. elegans* growth has also been examined in work not reported here, and worms exposed to water extracted from Lufa 2.2 soil with $100\,\mu g\,g^{-1}$ cadmium were significantly smaller than those exposed to water from soil with no cadmium. This experiment was carried out over 72 h, compared to the 24 h required for the stress response assay. The stress response may precede other physiological effects on the worm, and may provide advance warning of longer-term effects within a short assay timescale.

It would be useful to establish the extent to which metals found in soil after sewage sludge application change the soil biology. A reduction in soil microbial biomass on such sites is well documented (Brookes and McGrath 1984), which would, of course, have considerable effects on other groups of organisms. Nematodes are relatively under-represented in single species ecotoxiological tests (Maltby and Calow 1989). The ecology of other soil nematodes has also been proposed as a tool for indicating environmental disturbance, both at the level of changes in the patterns of species composition (Bongers 1990) and intra-specific changes (Kammenga 1995). These methodologies could provide a good benchmark for further assessment of the use of transgenic strains of *C. elegans* as biomonitors.

The heat shock or stress response represents a short-term cellular response to protein-damaging events that may be indicative of effects at higher levels of organization. It is proposed that a rapid microplate soil-toxicity assay could be developed based on the PC72 strain of *C. elegans*. This test would be able to contribute to the environmental diagnostics and biomonitoring of soils, and therefore have a role to play in forecasting the ecotoxicological effects of contaminants in soil.

REFERENCES

Alloway BJ (1990) *Heavy Metals in Soils*. Blackie, Glasgow, UK.

Bongers T (1990) The maturity index: an ecological measure of environmental disturbance based on nematode species composition. *Oecologica*, **83**, 14–19.

Brookes PC and McGrath SP (1984) Effects of metal toxicity on the size of the soil biomass. *Journal of Soil Science*, **35**, 341–346.

Calamari D and Alabaster JS (1980) An approach to theoretical models in evaluating the effects of mixtures of toxicants in the aquatic environment. *Chemosphere*, **9**, 533–538.

Chanmugathus P and Bollag JM (1987) Microbial role in immobilization and subsequent mobilization of cadmium in soil suspensions. *Soil Science Society of America Journal*, **51**, 1184–1191.

Daniells C, Duce I, Thomas D, Sewell P, Tattersall J and de Pomerai D (1998) Transgenic nematodes as biomonitors of microwave induced stress. *Mutation Research*, **9**, 55–64.

Dennis JL, Mutwakil MHAZ, Lowe KC and de Pomerai DI (1997) Effects of metal ions in combination with a non-ionic surfactant on stress responses in a transgenic nematode. *Aquatic Toxicology*, **40**, 37–50.

de Pomerai DI (1996) Heat-shock proteins as biomarkers of pollution. *Human and Experimental Toxicology*, **15**, 279–285.

Doelman P, Nieboer G, Schrooten J and Visser M (1984) Antagonistic and synergistic toxic effects of Pb and Cd in a simple foodchain: nematodes feeding on bacteria or fungi. *Bulletin of Environmental Contamination and Toxicology*, **32**, 717–723.

Donkin SG and Dusenbery DB (1993) Soil toxicity test using the nematode *Caenorhabditis elegans* and an effective method of recovery. *Archives of Environmental Contamination and Toxicology*, **25**, 145–151.

Elliot HA, Liberati MR and Huang CP (1986) Competitive adsorption of heavy metals by soils. *Journal of Environmental Quality*, **15**, 214–219.

Fire A (1986) Integrative transformation of *C. elegans*. *EMBO Journal*, **5**, 2673–2680.

Gaugler R, Wilson M and Shearer P (1997) Field release and environmental fate of a transgenic entomopathogenic nematode. *Biological Control*, **9**, 75–80.

Guven K, Duce JA and de Pomerai DI (1994) Evaluation of a stress-inducible transgenic nematode strain for rapid aquatic toxicity testing. *Aquatic Toxicology*, **29**, 119–137.

Guven K and de Pomerai DI (1995) Differential expression of HSP70 proteins in response to heat and cadmium in *Caenorhabditis elegans. Journal of Thermal Biology*, **20**, 355–363.

Guven K, Duce JA and de Pomerai DI (1995) Calcium moderation of cadmium stress explored using a stress-inducible transgenic strain of *Caenorhabditis elegans. Comparative Biochemistry and Physiology*, **110c**, 61–70.

Jones D, Stringham EG, Babich SL and Candido EPM (1996) Transgenic strains of the nematode *C.elegans* in biomonitoring and toxicology: effects of captan and related compounds on the stress response. *Toxicology*, **109**, 119–127.

Kammenga JE (1995) Phenotypic plasticity and fitness consequences in nematodes exposed to toxicants. PhD thesis, Department of Nematology, Wageningen Agricultural University, the Netherlands.

Lowry OH, Rosebrough N, Farr A and Randall R (1951) Protein measurements with the Folin-phenol reagent. *Journal of Biological Chemistry*, **193**, 165–175.

Maltby L and Calow, P (1989) The application of bioassays in the resolution of environmental problems; past, present and future. *Hydrobiologica*, **188/189**, 65–76.

Mutwakil MHAZ, Reader JP, Holdich DM, Smithurst PR, Candido EPM, Jones D and de Pomerai DI (1997) Use of stress-inducible transgenic nematodes as biomarkers of heavy metal pollution in water samples from an English river system. *Archives of Environmental Contamination and Toxicology*, **32**, 146–153.

Nowell MA, Wardlaw A, de Pomerai DI and Pritchard D (1997) The measurement of immunological stress in nematodes. *Journal of Helminthology*, **71**, 119–123.

Posthuma L, Baerselman R, van Veen RPM and Dirven-van Breemen EM (1997) Single and joint effects of copper and zinc on reproduction of *Enchytraeus crypticus* in relation to sorption of metals in soils. *Ecotoxicology and Environmental Safety*, **38**, 108–121.

Power RS, David HE, Mutwakil MHAZ, Fletcher K, Daniells C, Nowell MA, Dennis JL, Martinelli A, Wiseman R, Wharf E and de Pomerai DI (1998) Stress-inducible transgenic nematodes as biomonitors of soil and water pollution. *Journal of Biosciences*, **23**, 101–114.

Power RS and de Pomerai DI (1999) Effect of single and paired metal inputs on a stress-inducible transgenic nematode. *Archives of Environmental Contamination and Toxicology*, **37**, 503–511.

Sanders B (1990) Stress proteins: potential as multitiered biomarkers. In *Biomarkers of Environmental Contamination*, McCarthey JF and Shugart LR (eds), Lewis, Boca Raton, FL, USA.

Sanders B (1993) Stress proteins in aquatic organisms: an environmental perspective. *Critical Reviews in Toxicology*, **23**, 49–75.

Stringham EE, Dixon DK, Jones D and Candido EPM (1992) Temporal and spatial patterns of the small heat shock (*hsp*16) genes in transgenic *Caenorhabditis elegans. Molecular Biology of the Cell*, **3**, 221–233.

Stringham EE and Candido EPM (1993) Targeted single-cell induction of gene products in *Caenorhabditis elegans*: a new tool for developmental studies. *Journal of Experimental Zoology*, **266**, 227–233.

Stringham EE and Candido EPM (1994) Transgenic *hsp* 16-*lacZ* stains of the soil nematode *Caenorhabditis elegans* as biological monitors of environmental stress. *Environmental Toxicology and Chemistry*, **13**, 1211–1220.

Traunsperger W, Haitzer M, Höss S, Beier S, Ahlf W and Steinberg C (1997) Ecotoxicological assessment of aquatic sediments with *Caenorhabditis elegans* (Nematoda)—a method for testing liquid medium and whole sediment samples. *Environmental Toxicology and Chemistry*, **16**, 245–250.

van Gestel CAM and Hensbergen PJ (1997) Interaction of Cd and Zn toxicity for *Folsomia candida* Willem (Collembola: Isotomidae) in relation to bioavailability in soil. *Environmental Toxicology and Chemistry*, **16**, 1177–1186.

Williams PL and Dusenbery DB (1990) Aquatic toxicity testing using the nematode *Caenorhabditis elegans. Environmental Toxicology and Chemistry*, **9**, 1285–1290.

Can Animal Behaviour Predict Population Level Effects?

ERIK BAATRUP, MARK BAYLEY, FREDDY F. SØRENSEN AND GUNNAR TOFT

Institute of Biological Sciences, University of Aarhus, Denmark

12.1 INTRODUCTION

Forecasting the deleterious effects of existing and new chemicals presents one of the most challenging tasks in ecotoxicological research. By developing strong predictive tools we will hopefully, at least to a certain extent, be able to prevent instead of having to repair the effects of pollutants on living organisms. An important intermediate aim is the identification and development of biomarkers to assess the biological impact of chemical stressors. The scientific value and applicability of such biomarkers depend largely on which level of biological organization the measurements are made and how closely the responses are related to effects at both lower and higher levels (Depledge 1994). Biomarkers at the molecular level (Stegeman *et al.* 1992) respond rapidly and are often specific to a particular group of chemicals, and as such may be used to forecast the degree of exposure and absorption of chemicals. Unfortunately, these biochemical biomarkers have a limited value in assessing the consequences of the measured exposure for the organism (Mayer *et al.* 1992). At the population (Takamura *et al.* 1991) and community (Brock *et al.* 1992) levels, measured responses have strong ecological relevance, but in most cases limited predictive utility, because the damage has already occurred by the time it can be detected. Accordingly, biomarker responses at the organism level or below seem to have the broadest predictive applicability.

Quantitative analysis of animal behaviour may provide the functional link between the sub-organismal disturbances of chemical stress and population levels effects (Little and Finger 1990, Little *et al.* 1990, Scherer 1992, Baatrup and Bayley 1998). Changes in animal behaviour have a central position in the flow of pollutant-induced effects through the biological hierarchy of complexity with links in both directions (Figure 12.1). On the one hand, pollutant-induced changes of animal behaviour are intrinsically linked to impaired processes within the animal. Thus, animal locomotor behaviour has proven to be a sensitive biomarker of chemical stress in a number of animal species (Sørensen *et al.* 1995, Jensen *et al.* 1997) with responses persisting long after exposure has

Forecasting the Environmental Fate and Effects of Chemicals. Edited by Philip S. Rainbow, Steve P. Hopkin and Mark Crane.
© 2001 John Wiley & Sons Ltd

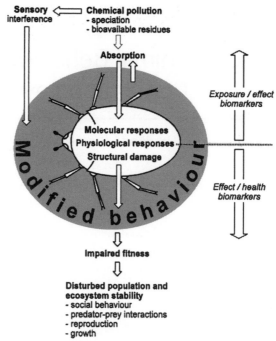

Figure 12.1 When a chemical is released into the environment, a chain of events through the levels of biological organization is initiated. The parent compound or residues will be absorbed by the animal from the food or across outer surfaces. The initial effects will always be at the molecular level, and if the chemical is not neutralized by the organism, these molecular changes may have deleterious consequences for cell function and survival. If many or critical cells are affected, alterations in vital physiological processes will weaken the condition of the animal. This will almost certainly be expressed in an altered behaviour. For animals in the wild, even minor changes in behaviour may reduce their fitness through for example their ability to obtain food, avoid predation and successfully reproduce.

ceased (Bayley 1995) and across changes in life stage (Bayley *et al.* 1995). On the other hand, the behaviour of an animal is involved in such vital life processes as feeding, predator avoidance, reproduction and migration, all events of great importance to the individual and population maintenance.

Although alterations in animal behaviour are, as stated by Little (1990), intuitively linked to effects at the population level, direct evidence of such mechanistic links still remain to be established. Below, we have selected three recent studies, which illustrate the potential applicability of behavioural analyses in forecasting population level effects.

12.2 ELUCIDATING ANIMAL BEHAVIOUR FROM DIGITAL IMAGES

Computer-aided video tracking is employed in all our studies of animal behaviour. Terrestrial animals are measured in test arenas walking on their natural or artificial substrates, while adequate size aquaria are used for aquatic animals. A camera transmits an image of the scenario to a frame grabber inside a computer where the captured images are digitized at a high rate. Within each digital image, the animals are identified on the basis of preassigned colour, size and shape criteria. For the various behavioural analyses, we have developed specialized tracking systems. In studies addressing only locomotor behaviour, the positions of the animal centroids are stored as time series of Cartesian coordinates from which a number of locomotor parameters can be extracted. In cases where more complex behavioural patterns are analysed, such as the guppy sexual behaviour described below, each animal must be characterized in more detail in the digital images in order to describe its shape, together with its position and orientation relative to other individuals. In such cases, the pixel mass representing each animal is converted into one or several systems of coordinates in order to recognize head and tail, the shape of the animal and its position in other animals' systems of coordinates. Selected components of complex behaviours can then be analysed individually or in combination.

12.3.1   CHEMICAL STRESS AND COLLEMBOLAN AGGREGATION BEHAVIOUR

Many collembolan species form dense aggregations in the field and in laboratory cultures. The search for patches with appropriate conditions, such as moisture and food availability, partly explains collembolan aggregation behaviour (Usher and Hide 1975). However, there is also convincing evidence that Collembola are mutually attracted by pheromones (Verhoef *et al.* 1977, Mertens *et al.* 1979, Leonard and Bradbury 1984). Pheromone-led aggregation may play several biological roles of importance for population maintenance. Firstly, in sexually reproducing species, aggregation will increase the chances of successful collection of spermatophores by females. Secondly, non-sexual attraction around rich and high-quality food may be beneficial to the population and thirdly, closely aggregated Collembola may create their own microclimate, which may reduce the risk of desiccation (Hopkin 1997).

At our laboratory we have studied the distribution patterns of the collembolan *Folsomia candida* in Petri dishes following a sub-lethal exposure to the organophosphorous pesticide dimethoate. Following exposure, the nearest-neighbour distances of 40 individuals were assessed daily for eight consecutive days by image analysis in 10 replicates of exposed and control populations. The average nearest-neighbour distance was then calculated and presented as the ratio to the mean of the nearest-neighbour distance of a randomly distributed population, obtained by computer simulation. The control Collembola, though not densely aggregated, were clearly attracted to each other (ratio of 0.68), whereas the exposed groups were nearly randomly dispersed (ratio of 0.95). A disturbed sensation or responsiveness to aggregation pheromones, caused by the toxic action of the pesticide, could account for this increased dispersion in exposed collembolan populations. To test this theory, one half of small test

Figure 12.2   Aggregation in Collembola is partly mediated through pheromone attraction. The left figure shows the track of an individual walking in an arena, where the top half of the area was conditioned beforehand for 24 h by 30 conspecifics. The right track was made by a collembolan slightly intoxicated with the organophosphate dimethoate.

arenas were conditioned for 24 h by 30 conspecifics. Following conditioning, a single collembolan was placed in each arena and its moving pattern measured for three hours. Typical tracks for control and dimethoate-exposed Collembola are shown in Figure 12.2. The control Collembola clearly preferred the conditioned half of the arena, where they spent less time in activity, moved with lower velocities, had a higher turning rate and frequency of stopping, and a lower right/left turning bias, when compared with the unconditioned half. Overall, this moving pattern was unchanged in exposed Collembola. However, in comparison with the controls, individuals exposed to dimethoate frequented the unconditioned section five times more often and showed a significantly higher locomotor activity. Thus, they walked more than three times the distance of the controls, employing much higher movement velocities.

Apparently, the exposed Collembola were still able to perceive traces of conspecifics, but failed to respond normally to the presence or absence of stimuli and the transition between them. This may be a contributing factor to the observed effects of dimethoate on aggregation behaviour. Together, these two experiments illustrate by example the importance of behavioural changes in vital population events and as such the potential predictive value of unbiased behavioural measurements.

12.3.2   ENDOCRINE DISRUPTORS AFFECT GUPPY SEXUAL BEHAVIOUR AND REPRODUCTION

The second example presented here emphasizes that behavioural disruption is only one of several mechanisms underlying the response of a population to chemical stress.

Recent findings indicate that widely used chemicals with endocrine disrupting properties, are disturbing development and maintenance of reproductive function. Falling sperm counts and increased instances of testicular deformations are observed in humans and are possibly caused by exposure to endocrine disrupting chemicals (Sharpe and Skakkebæk 1993). In fish, an increase in the yolk protein vitellogenin and reduced gonadosomatic index (GSI) are seen in laboratory experiments with male trout exposed to alkyl phenolic compounds (Jobling et al. 1996). Field studies with caged male trout exposed to sewage effluents have shown similar effects on GSI and vitellogenin (Harries et al. 1997).

The potential threat to human and animal reproduction posed by the presence of xenoestrogens in the environment has been met by an intensive effort to identify and develop fast and reliable biomarkers of both exposure and effects of suspected endocrine disrupting chemicals. *In vitro* methods have been developed specifically to detect the presence of oestrogenic compounds. These methods include recombinant yeast cells with the human oestrogen receptor (Routledge and Sumpter 1996) and the oestrogen-inducible MCF-7 breast cancer line (White et al. 1994). Similarly, vitellogenesis is a sensitive *in vivo* biomarker of oestrogen and xenoestrogen exposure (Sumpter and Jobling 1995). While these methodologies have proven highly sensitive to xenoestrogenic activity of chemicals and particularly in the case of vitellogenesis, have been central in showing the presence of these chemicals in the environment (e.g. Jobling and Sumpter 1993, Harries et al. 1996), they do not provide evidence of effects on animal reproduction itself. To bridge this gap, biomarkers at higher levels of biological organization are needed.

The reproductive success of animals depends strongly on an appropriate sexual behaviour. When released by physical or chemical stimuli, adequate hormone levels trigger courtship and mating behaviour, which are prerequisites of fertilization. An imbalance in these natural hormones inevitably results in abnormal sexual behaviour and consequently, impaired reproduction. Similar dysfunctions can be caused by hormone-mimicking substances absorbed from the environment (Bryan et al. 1989, Colgan et al. 1982, Matthiessen and Logan 1984, White et al. 1983, Whitten et al. 1995).

In three experiments we have considered the effects of the natural oestrogen 17β-estradiol (E2) and the xenoestrogen 4-*tert*-octylphenol (OP) on male guppy sperm count (cellular level), coloration (organ level) courtship behaviour (individual level) and on the sexual maturation of juveniles and their later reproductive capability as adults (population level). The guppy (*Poecilia reticulata*) was chosen as experimental organism for several reasons. Most importantly, the sexual display of the male towards the female, known as the **sigmoid display**, is easily distinguishable from the rest of the behavioural repertoire (Houde 1997), is perpetually performed by the male, and can be quantified automatically by a computer. Also, the guppy is a viviparous continuously breeding animal with gonad size, sperm count and reproductive output independent of season (Houde 1997).

In one experiment, adult male guppies were continuously exposed to nominal concentrations of 150 μg l^{-1} OP and 10 μg l^{-1} E2 for one month (Bayley et al. 1999).

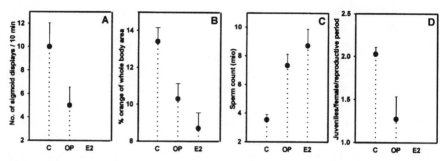

Figure 12.3 Sexual behaviour (A), male coloration (B), sperm count (C) and reproduction (D) of guppies treated with 4-*tert*-octylphenol (OP) and 17β-estradiol (E2). Values are presented as means ($\forall$ standard error). (A) adult males exposed to 150 μg l^{-1} OP or 10 μg l^{-1} E2. (B and C) adult males exposed to 300 μg l^{-1} OP or 1 μg l^{-1} E2. (D) juveniles exposed to 10 μg l^{-1} OP or 5 μg l^{-1} E2.

When compared with a control group, these treatments caused a significant reduction in the number of sigmoid displays in the OP-exposed group and a total extinction of sexual behaviour in the E2-exposed group (Figure 12.3A). Subsequent analyses of the fish spontaneous locomotor behaviour in the absence of females indicated that the behavioural changes were not merely a general toxic response.

The dose dependency of OP and E2 on body coloration and sperm count was measured in a subsequent study. The orange spots of the male fish, which play an important role in the female's sexual selection of males, were significantly smaller and paler in treated animals (Figure 12.3B). The size of orange areas was negatively correlated with the OP and E2 concentrations in the water. Further, a significant increase in sperm count was observed in OP and E2 treated adult male guppies (Figure 12.3C). The maximum sperm count in OP treated animals appeared at the medium concentrations of 300 μg l^{-1}, while E2 induced the maximum sperm count at 1 μg l^{-1}, the highest dose tested. In the third experiment, the effects of E2 and OP on juvenile development and reproductive output of these animals as adults were quantified. Groups of juvenile guppies were exposed to 5 μg l^{-1} E2 and 10 μg l^{-1} OP, respectively, from day 0 to day 21 after birth. E2 caused all juveniles to develop into females, while OP delayed male development, but did not significantly change the sex ratio. After sexual maturation, the reproductive output of these guppies was determined as the number of neonates during five reproductive periods. The average number of newborns per female per reproductive period was significantly smaller in the OP exposed group compared to the control group (Figure 12.3D). Obviously, no juveniles were born in the E2 group where all males were sex reversed.

These studies on guppies clearly show that 17β-estradiol and 4-*tert*-octylphenol exert effects on several levels of biological organization ranging from sperm count at the cellular level, coloration at the organ level, behaviour and development at the organism level, to changes in sex ratio and impaired reproduction at the population level. The reduced reproductive output of the guppy population exposed to OP undoubtedly resulted from irreversible (structural) modifications of the individual (females or males?) during the time of exposure early in life. Direct mechanistic relationships between the measured effects at the different levels were not established, but clearly support the existence of such links. Farr (1980) demonstrated that the reproductive success in guppies depends on the male sexual behaviour, and it is well documented that the female sexual selection depends on the size (Kodric-Brown 1985, Houde 1987) and intensity (Houde and Torio 1992) of the male coloration. Similarly, a positive correlation between sperm count and sexual behaviour has been observed in natural guppy populations (Matthews *et al.* 1997). The high sperm counts in the E2 and OP exposed groups of the present study might be due to precocious conversion of early spermatogenic elements into mature sperm by oestrogens as proposed by Berkowitz (1941), or caused by fluid reabsorption from the seminal fluid as seen in the mouse (Hess *et al.* 1997).

12.3.3 POLLUTED SOIL ALTERS WOODLOUSE LOCOMOTOR BEHAVIOUR

The final example demonstrates that changes in animal behaviour are not restricted to controlled laboratory conditions but may also occur in response to chemical stress in the field. In their natural environment, animals are constantly affected by varying physical conditions such as humidity and temperature, and at polluted sites they are almost always exposed to complex chemical mixtures. It was our hypothesis, that alterations in behaviour express an integrated response to the animal's internal status, including chemical stress. We therefore wished to test whether animal populations from polluted locations differ in their locomotor behaviour when compared with populations from clean environments. For this purpose, woodlice (Crustacea, Isopoda) are ideally suited as indicator organisms in the terrestrial environment because as saprophages, they directly encounter and consume polluted plant material, have a wide global distribution and are common inhabitants of both urban and rural habitats (Hopkin *et al.* 1986). Furthermore, because they live in a highly unstable environment, woodlice depend on simple behavioural reactions in response to unfavourable conditions. Several studies have emphasized the importance of behavioural adaptations to avoid water loss (Warburg 1987). Also, locomotion is clearly a prerequisite of predator avoidance, food seeking, and reproduction (Sutton 1972), all life processes of decisive importance to population stability. Consequently, measured alterations in the behaviour of individual woodlice are likely to indicate impairment of population health.

Sørensen *et al.* (1997) studied woodlice from a polluted English woodland. During a severe fire at a factory in 1991, 40 t of molten plastic flowed into the site covering a distinct area of about 2500 m². Five years later, the layer of pyrolysed plastic was covered with soil and leaf-litter severely contaminated with a complex mixture of heavy metals and various organic compounds. Woodlice were collected along a transect from the central part of the polluted area to an uncontaminated site 100 m from the pollution. The woodlice collected directly from the pyrolysed plastic layer spent significantly less time in locomotor activity, resulting in a walked distance less than 50% of that of the individuals collected at the reference site. Furthermore, a disturbed turning behaviour and a less continuous moving pattern was found for the woodlice collected on the pyrolysed plastic. This aberrant behaviour coincided with significant elevated body concentrations of heavy metals and lowered protein and glycogen levels. Although this study indicated that locomotor behaviour could reveal effects in populations of animals exposed to chemical mixtures in the field, it left fundamental questions unanswered. In particular, what is the variation in locomotor behaviour **between** the reference sites with which the animals from contaminated areas are to be compared? Further, which elements in locomotor behaviour are the most efficient indicators of chemical stress?

These questions were addressed in a study in Denmark centred around a recently closed iron foundry (Bayley et al. 1997). The question of variability in the behaviour of woodlice from distinct reference sites was approached by adopting an asymmetrical experimental design, where the behaviour of woodlice from the polluted site was compared with animals from **four** unpolluted reference sites. Those elements of the locomotor behaviour which are the best indicators of chemical stress were identified using linear discriminant analysis. This statistical technique combines the locomotor components into a linear equation which describes the function that best separates the groups. If such discriminant models have a sufficiently solid foundation, then they have a considerable predictive value and can be used to test animals from new localities. The analysis showed that the four control populations were remarkably similar in their locomotor behaviour despite their geographic separation of more than 300 km. The discriminant function also showed that the foundry population differed significantly from the controls in terms of walked distance and movement velocities, whereas other locomotor parameters contributed little to group separation. Again, the deviant behaviour was accompanied by elevated hepatopancreas and whole-body metal concentrations. However, in addition to metals, the animals at the foundry were also exposed to a cocktail of other environmental contaminants.

In a subsequent study on the island of Fynen (Bayley and Baatrup unpublished), the experimental design was expanded to include four clean reference sites separated by 50–100 km and three sites with different types of pollution. An industrial rubbish dump, closed since the 1960s, was polluted almost exclusively by zinc, cadmium and lead, while the soil of a former coal-gas works was heavily contaminated with cyanide and a range of organic pollutants. The third location was a partially renovated tar-asphalt works where soil was contaminated with PAHs. The results of this study confirmed that the locomotor behaviour of control populations was remarkably similar, and therefore independent of geographical location, at least on the scale available in Denmark (Figure 12.4). The locomotor behaviour of the woodlice collected from two of the polluted sites (the former rubbish dump and coal-gas works) was again found to be significantly different from all controls, whereas the behaviour of the animals from the PAH-contaminated soil was not (Figure 12.4). It was not possible to discriminate the three very different pollution types from each other solely on the basis of the woodlouse locomotor behaviour. Possibly such a distinction can be achieved by including other biomarkers in the discriminant model.

12.4 CONCLUSIONS

In conclusion, although none of the cited studies provide evidence that behavioural disturbances influence population stability, they do indicate the predictive value of behaviour analyses in forecasting such effects. It is difficult to think of another biomarker, which is so easy to measure and at the same time integrates metabolic, neural and endocrine disturbances. However, before the links between behavioural changes and effects at the population and community levels can be established, more basic research is needed. Firstly, more studies providing evidence that pollutants in the field cause behavioural changes are required. Also, we need to be more confident that observed behavioural differences between reference and exposed populations are in fact due to pollutants and are not an artefact of some unrecognised factor at a single site. One way to strengthen conclusions is to use multiple reference sites or to establish an extensive database on the 'normal variability' of biomarker

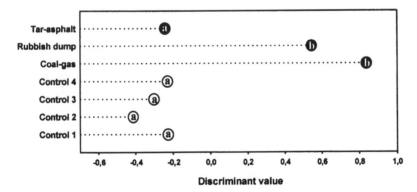

Figure 12.4 The locomotor behaviour of the woodlouse *Oniscus asellus* is modulated by its chemical environment. Employing discriminant analysis, all measured elements in the locomotor behaviour are weighed, making it possible to separate woodice living in polluted sites from individuals collected from unpolluted sites. The average discriminant value of 30 woodlice from each of four reference sites (controls 1–4) were significantly different from the values of woodlice collected at a former industrial rubbish dump and a closed coal-gas works, but not different from animals collected from a renovated tar-asphalt works. Discriminant values marked by the same letter are not significantly different ($p < 0.05$) from each other (Tukey's test).

responses (Depledge 1994, McCarthy 1994, Bayley *et al.* 1997). Secondly, we need more studies examining the ecological significance of chemically induced behavioural deviations. If the ecological consequences of the behavioural changes are well understood, our ability to predict adverse population effects in the field will be greatly improved. An important step in the development of 'behaviour biomarkers' is therefore the establishment of links to effects at higher levels of biological organization. We definitely need proof of what we assume to be intuitively obvious, namely that behavioural changes contribute to reduced Darwinian fitness (e.g. survival, growth and reproduction).

REFERENCES

Baatrup E and Bayley M (1998) Animal locomotor behaviour as a health biomarker of chemical stress. In *Diversification in Toxicology—Man and Environment*, Seiler JP, Autrup JL and Autrup H (eds), *Archives of Toxicology* (Suppl. 20), Springer, Berlin, pp. 163–178.

Bayley M (1995) Prolonged effects of the insecticide dimethoate on locomotor behaviour in the woodlouse, *Porcellio scaber* Latr. (Isopoda). *Ecotoxicology*, **4**, 79–90.

Bayley M, Baatrup E, Heimbach U and Bjerregaard P (1995) Elevated copper levels during larval development cause altered locomotor behavior in the adult carabid beetle *Pterostichus cupreus* L. (Coleoptera: Carabidae). *Ecotoxicology and Environmental Safety*, **32**, 166–170.

Bayley M, Baatrup E and Bjerregaard P (1997) Woodlouse locomotor behavior in the assessment of clean and contaminated field sites. *Environmental Toxicology and Chemistry*, **16**, 2309–2314.

Bayley M, Nielsen JR and Baatrup E (1999) Guppy sexual behaviour as an effect biomarker of estrogen mimics. *Ecotoxicology and Environmental Safety*, **43**, 68–73.

Berkowitz P (1941) The effects of estrogenic substances in the fish (*Lebeistes reticulatus*). *Journal of Experimental Zoology*, **87**, 233–243.

Brock TCM, van den Bogaert M, Bos AR, Breukelen SWF, Reiche R, Terwoert J, Suykerbuyk REM and Roijackers RMM (1992) Fate and effects of the insecticide dursban 4E in indoor elodea-dominated and marcrophyte-free freshwater model ecosystems: II Secondary effects on community structure. *Archives of Environmental Contamination and Toxicology*, **23**, 391–409.

Bryan TE, Gildersleeve RP and Wiard RP (1989) Exposure of Japanese quail embryos to *o,p'*-DDT has long-term effects on reproductive behaviors, hematology, and feather morphology. *Teratology*, **39**, 525–535.

Colgan PW, Cross JA and Johansen PH (1982) Guppy behaviour during exposure to a sub-lethal concentration of phenol. *Bulletin of Environmental Contamination and Toxicology*, **28**, 20–27.

Depledge MH (1994) The rational basis for the use of biomarkers as ecotoxicological tools. In *Nondestructive Biomarkers in Vertebrates*, Fossi MC and Leonzio C (eds), Lewis, Boca Raton, FL, pp. 271–297.

Farr JA (1980) Social behaviour patterns as determinants of reproductive succes in the guppy, *Poecilia reticulata* Peters (Pisces: Poeciliidae). An experimental study of the effects of intermale competition, female choice, and sexual selection. *Behaviour*, **74**, 38–91.

Harries JE, Sheahan DA, Jobling S, Matthiessen P, Neall P, Routledge EJ, Rycroft R, Sumpter JP and Tylor T (1996) A survey of estrogenic activity in United Kingdom inland waters. *Environmental Toxicology and Chemistry*, **15**, 1993–2002.

Harries JE, Sheahan DA, Jobling S, Matthiessen P, Neall P, Sumpter JP, Tylor T and Zaman N (1997) Estrogenic activity in five United Kingdom rivers detected by measurement of vitellogenesis in caged male trout. *Environmental Toxicology and Chemistry*, **16**, 534–542.

Hess RA, Bunick D, Lee KH, Bahr J, Taylor JA, Korach KS and Lubahn DB (1997) A role for oestrogens in the male reproductive system. *Nature*, **390**, 509–512.

Hopkin SP (1997) *Biology of the Springtails (Insecta: Collembola)*. Oxford University Press, Oxford.

Hopkin SP, Hardisty GN and Martin M.H (1986) The woodlouse *Porcellio scaber* as a 'biological indicator' of zinc, cadmium, lead and copper pollution. *Environmental Pollution*, 11B, 271–290.

Houde AE (1987) Mate choice based upon naturally occurring colour-pattern variation in a guppy population. *Evolution*, **41**, 1–10.

Houde AE (1997) *Sex, Color, and Mate Choice in Guppies*. Monographs in Behavior and Ecology. Krebs JR and Clutton-Brock T (eds), Princeton University Press, Princeton, NJ.

Houde AE and Torio J (1992) Effects of parasitic infection on male colour pattern and male choice in guppies. *Behavioural Ecology*, **3**, 346–351.

Jensen CS, Garsdal L and Baatrup E (1997) Acetylcholinesterase inhibition and altered locomotor behaviour in the carabid beetle *Pterostichus cupreus*. A linkage between biomarkers at two levels of biological complexity. *Environmental Toxicology and Chemistry*, **16**, 1727–1732.

Jobling S and Sumpter JP (1993) Detergent components in sewage effluent are weakly oestrogenic to fish: an in vitro study using rainbow trout (*Oncorhynchus mykiss*) hepatocytes. *Aquatic Toxicology*, **27**, 361–372.

Jobling S, Sheahan D, Osborne JA, Matthiessen P and Sumpter JP (1996) Inhibition of testicular growth in rainbow trout (*Oncorhynchus mykiss*) exposed to estrogenic alkylphenolic chemicals. *Environmental Toxicology and Chemistry*, **15**, 194–202.

Kodric-Brown A (1985) Female preference and sexual selection for male coloration in the guppy (*Poecilia reticulata*). *Behavioural Ecology and Sociobiology*, **17**, 199–205.

Leonard MA and Bradbury PC (1984) Aggregative behaviour in *Folsomia candida* (Collembola: Isotomidae) with respect to previous conditioning. *Pedobiologia*, **26**, 369–372.

Little EE (1990) Behavioral toxicology: Stimulating challenges for a growing discipline. *Environmental Toxicology and Chemistry*, **9**, 1–2.

Little EE, Archeski RD, Flerov BA and Kozlovskaya VI (1990) Behavioural indicators of sublethal toxicity in rainbow trout. *Archives of Environmental Contamination and Toxicology*, **19**, 380–385.

Little EE and Finger SE (1990) Swimming behaviour as an indicator of sublethal toxicity in fish. *Environmental Toxicology and Chemistry*, **9**, 13–19.

Matthews IM, Evans JP and Magurran AE (1997) Male display rate reveals ejaculate characteristics in the Trinidadian guppy *Poecilia reticulata*. *Proceedings of the Royal Society London*, Series B, **264**, 695–700.

Matthiessen P and Logan JWM (1984) Low concentration effects of endosulfan insecticides on reproductive behaviour in the tropical cichlid fish *Sarotherodon mossambicus*. *Bulletin of Environmental Contamination and Toxicology*, **33**, 575–583.

Mayer FL, Versteeg DJ, McKee MJ, Folmar LC, Graney RL, McCurme DC and Rattner BA (1992) Physiological and nonspecific biomarkers. In *Biomarkers: Biochemical, Physiological, and Histological Markers of Anthropogenic Stress*, Huggett RJ (ed.), Lewis, Chelsea, pp. 5–85.

McCarthy JF (1994) The future of nondestructive biomarkers. In *Nondestructive Biomarkers in Vertebrates*, Fossi MC and Leonzio C (eds), Lewis, Boca Raton, FL, pp. 313–324.

Mertens J, Blancquaert JP and Bourgoignie R (1979) Aggregation pheromone in *Orchella cincta* (collembola). *Revue d'Écologie et de Biologie du Sol*, **16**, 441–447.

Routledge EJ and Sumpter JP (1996) Estrogenic activity of surfactants and some of their degradation products assessed using a recombinant yeast screen. *Environmental Toxicology, and Chemistry*, **15**, 241–248.

Scherer E (1992) Behavioural responses as indicators of environmental alterations: Approaches, results, developments. *Journal of Applied Ichthyology*, **8**, 122–131.

Sharpe RM and Skakkebæk NE (1993) Are oestrogens involved in falling sperm counts and disorders of the male reproductive tract? *Lancet*, **341**, 1392–1395.

Sørensen FF, Bayley M and Baatrup E (1995) The effects of sublethal dimethoate exposure on the locomotor behavior of the collembolan *Folsomia candida* (Isotomidae). *Environmental Toxicology and Chemistry*, **14**, 1587–1590.

Sørensen FF, Weeks JM and Baatrup E (1997) Altered locomotor behaviour in woodlouse (*Oniscus asellus* L.) collected at a polluted site. *Environmental Toxicology and Chemistry*, **16**, 685–690.

Stegeman JJ, Brouwer M, Di Giulio RT, Förlin L, Fowler BA, Sanders BM and van Veld PA (1992) Molecular responses to environmental contamination: Enzyme and protein systems as indicators of chemical exposure and effects. In *Biomarkers: Biochemical, Physiological, and Histological Markers of Anthropogenic Stress*, Hugget RJ (ed.), Lewis, Cheasea, pp. 235–335.

Sumpter JP and Jobling S (1995) Vitellogenesis as a biomarker for estrogenic contamination of the aquatic environment. *Environmental Health Perspectives*, **103**, 173–178.

Sutton SL (1972) *Woodlice*. Ginn, London, UK.

Takamura K, Hatakeyama S and Shiraishi H (1991) Odonate larvae as an indicator of pesticide contamination. *Applied Entomology and Zoology*, **26**, 321–326.

Usher MB and Hide M (1975) Studies on populations of *Folsomia candida* (Insecta: Collembola): causes of aggregation. *Pedobiologia*, **15**, 276–283.

Verhoef HA, Nagelkerke CJ and Joosse ENG (1977) Aggregation pheromones in collembola. *Journal of Insect Physiology*, **23**, 1009–1013.

Warburg MR (1987) Isopods and their terrestrial environment. *Advances in Ecological Research*, **17**, 187–242.

White DH, Mitchell CA and Hill EF (1983) Parathion alters incubation behavior of laughing gulls. *Bulletin of Environmental Contamination and Toxicology*, **31**, 93–97.

White R, Jobling S, Hoare SA, Sumpter JP and Parker MG (1994) Environmentally persistent alkylphenolic compounds are estrogenic. *Endocrinology*, **135**, 175–182.

Whitten PL, Lewis C, Russel E and Naftolin F (1995) Phytoestrogen influences on the development of behaviour and gonadotropin function. *Proceedings of the Society for Experimental Biology and Medicine*, **208**, 82–86.

13

Forecasting Effects of Toxicants at the Community Level: Four Case Studies Comparing Observed Community Effects of Zinc with Forecasts from a Generic Ecotoxicological Risk Assessment Method

LEO POSTHUMA, TON SCHOUTEN, PATRICK VAN BEELEN AND MICHIEL RUTGERS

Laboratory for Ecotoxicology and Laboratory for Soil and Groundwater Research, RIVM — National Institute of Public Health and the Environment, Bilthoven, The Netherlands

13.1 INTRODUCTION

13.1.1 POLICY DOMAIN AND FRAMEWORK

Soil protection or soil remediation decisions are often based on benchmark concentrations for toxic compounds. In the domain of soil management policy, benchmark concentrations are usually considered as universally applicable total soil concentrations, that should discriminate between 'unaffected' and '(seriously) affected'. By not using location-specific data, they relate to ambient concentrations that are **potentially** hazardous for exposed communities.

Benchmark concentrations are usually derived on the basis of toxicity data, by means of a statistical extrapolation procedure, following the species sensitivity distribution (SSD) concept. This concept has a relatively long history, and has independently evolved in the United States and Europe (since 1976 (USEPA 1978) and 1985 (Kooijman 1987), respectively). Despite the use of the SSD concept for environmental legislation in various ways in different countries, the accuracy of the method has not often been investigated. A book (Posthuma,

With contributions from Dick Bakker, Trudie Crommentuijn, Herman Eijsackers, Olivier Klepper, Gerard Korthals, Jos Notenboom, Els Smit, Kees van Gestel, and Hans Vonk.

Traas and Suter, in prep.), reviewing the historical developments, the principles and characteristics and the various forms of usage of the SSD concept in environmental assessments is currently being prepared.

For ethical and operational reasons, the input data used in the SSD approach are mostly obtained under experimental conditions, testing separate species or microbial functions in artificially contaminated media. This has raised public and policy concerns on the meaning of benchmark concentrations (Hopkin 1993, van Straalen 1993, Chapman 1995a,b). Are they under- or overprotective when considering soil protection? Do they adequately lead to clean-up decisions when looking at soil remediation?

In this chapter, we investigate the ecological meaning of benchmark concentrations for soil that have been generated by a generic ecotoxicological risk assessment method (van Straalen and Denneman 1989). We investigate whether they truly relate to absence (protection) and potential presence (remediation) of adverse effects on biotic communities in the field.

13.1.2 SCIENTIFIC DOMAIN

To judge the predictive accuracy of benchmark concentrations for soil protection and remediation, the following scientific key issues are important:

1. Laboratory toxicity data: what is the ecological relevance of laboratory toxicity data, how should laboratory-to-field extrapolation be carried out to optimize their field relevance and which extrapolation uncertainties should be reduced?
2. Extrapolation to the community level: how should differences in species sensitivities be taken into account, and can statistical (or other) methods be used to make this extrapolation?
3. Field relevance of benchmark concentrations: how do they relate to adverse effects of contamination on community end points in field conditions?

This chapter presents experimental data that have been generated within a recent research programme focusing on the first and third question (Posthuma 1997, Posthuma *et al.* 1998b). With respect to the first issue, it was concluded that refinements can be made on the design and interpretation of laboratory toxicity tests, to improve their ecological relevance. However, current laboratory toxicity data compiled from literature cannot be adapted to the proposed improvements, since crucial test conditions are often not reported. Current benchmarks are based on imperfect input data, and thus, the analyses made in this chapter investigate whether **current** practice, with **current** imperfect toxicity data, yields ecologically meaningful benchmarks.

13.1.3 APPROACHES

To judge the ecological relevance of the benchmark derivation method in general, we have collected data on the adverse effects of a model compound on community end points in four case studies. Care has been taken that the benchmark concentrations for a compound were derived using a large range of laboratory toxicity data, to minimize the chance of bias introduced by limitations in the toxicological input data. In the chapter, emphasis is on case studies, and these are summarized and reviewed. Less attention is paid to the explanation of the risk assessment model. For further information the reader should refer to the literature. The choice of the risk assessment model was only a matter of (pragmatic) choice; the ecological relevance of other risk assessment models can also be investigated by comparison to true field effects.

13.2 FIELD OBSERVATIONS ON COMMUNITY RESPONSES TO METALS

13.2.1 CHOICE OF COMPOUNDS AND ORGANISM GROUPS

Four case studies were carried out to obtain field effect data. Zinc was chosen as non-degradable model toxicant. Although zinc is a micro-nutrient, zinc concentrations in the studies were so high that deficiency is unlikely. Nematodes and micro-organisms were chosen as model organisms, since they are intimately associated with the soil, they have short response times, and have a high within-taxon or functional variability. These characteristics render the organism groups practically and methodologically suitable to solve the issues at stake. Benchmarks have been separately calculated from toxicity data for soil invertebrates and microbial processes (see section 13.3), in order to address effects on structure and function of the community separately.

13.2.2 CHOICE OF SITES

Community toxicity end points were studied in an outdoor experimental field plot and in a metal contamination gradient in the field. Physico-chemical characteristics are summarized in Table 13.1.

1.   The experimental field plot consisted of 60 enclosures of $0.5\,m^2$ isolated by stainless steel plates. Each enclosure was filled with a sandy soil collected at an uncontaminated Dutch field site. Sand, containing shell particles, was added to improve soil structure. Some soil was set aside, for inoculating indigenous nematodes after artificial contamination of soil batches. In a completely randomized block design, different amounts of zinc were added (as $ZnCl_2$) to obtain initial zinc concentrations between 32 and $3200\,mg\,kg^{-1}$ dry wt soil, and control enclosures treated with water only. After these treatments, enclosure soils were inoculated with untreated soil samples containing the indigenous soil biota. It is noted: (1) that soil pH increased due to slow release of calcium from the shell particles in the added sand, (2) that differences between Zn treatments occurred for the factor soil pH (one unit lower at high zinc

Table 13.1 Characteristics of the soils used in the case studies (ref. 1 = Smit *et al.* 1997, refs 2 = Posthuma and Notenboom 1996, Smit and van Gestel 1996, Posthuma 1992)

	Experimental field plot (ref. 1)	Field gradient (refs 2)
Organic matter (%)	2.0	2.4–6.4
Clay content (%)	2.9	1.2–1.6
pH (KCl)	Initial, control 5.2	2.9–5.2
	After 22 months, control 7.1	
[Zn] mg kg^{-1} dry soil (total)	Control 32–3200	11–1800

concentrations) and (3) that zinc availability reduced over time due to leaching and increased sorption over time.

2. The field gradient is located in an area with sandy soils near Budel (the Netherlands). The site is contaminated by metals due to smelting activities between 1892 and 1973, and by the local deposition of metal-slag for road building. On a regional scale, a gradient of total metal concentrations can be recognized. On a detailed scale, however, total metal concentrations and metal availability may vary as a function of local historical events and soil characteristics.

13.2.3 STRUCTURE AND FUNCTION RELATED COMMUNITY END POINTS

The effects of metal exposure on the nematodes were investigated after extracting living nematodes from soil samples using Oostenbrink's elutration method. Extracted specimens were mounted on slides, and 150 individuals per sample were identified to the species level where possible, else to genus or family level. Density and diversity measures, among which density per species (or higher taxon), and the Shannon–Weaver index (see e.g. Lande 1996), were used as community end points.

The effects of metal exposure on the microbial communities were determined with two independent methods, focusing on the development of tolerance of microbial metabolic processes to zinc exposure. Increased tolerance is considered evidence for a selectively active (toxic) exposure. This so-called PICT (pollution-induced community tolerance, Blanck *et al.* 1988) approach can be applied to soil micro-organisms (Posthuma 1997). The method quantifies relative differences in metal tolerance of communities from different sites, using artificial exposure of the communities to the supposedly field-toxic compound. The degree of tolerance was used as the community end point.

13.2.4 CASE STUDY CHARACTERISTICS

1. **Case 1 (Schouten *et al.* 1998).** The effects of zinc on nematode communities in the experimental field plot were determined. Soil samples were taken 3, 10 and 22 months after inoculation. Samples were taken from five replicate enclosures at each zinc concentration.

2.   **Case 2 (Posthuma et al. 1998b).** The effects of metals on nematode communities along the field gradient were determined. Soil samples were taken in March 1993 at 10 sites with different total metal (Zn, Cu, Pb, Cd) concentrations.

3.   **Case 3.** The effects of zinc on microbial community tolerance in the experimental field plot were determined.

   - **Case 3a (van Beelen et al. 1998).** PICT was assessed after 19 months of exposure using tolerance changes of the acetate breakdown capacity in six out of ten zinc treatments. The soil samples were diluted ($2\,\mathrm{mg\,l^{-1}}$ in 20 mM tris buffer at pH 8) and incubated for 48 h at 20 °C. Interference by sorption of zinc to soil, or by growth of resistant micro-organisms during the toxicity test, was minimized in the procedure applied. To quantify the breakdown activity, 30 nM ^{14}C-labelled acetate was added, and the ^{14}C evolution rate was measured.

   - **Case 3b (Rutgers et al. 1998a).** PICT was assessed after 31 months of exposure using tolerance changes of 95 substrate breakdown activities in six out of the ten zinc treatments. Microbes were extracted in 1 mM bis-tris buffer pH 7. Soil particles were removed by centrifugation. A simultaneous study on PICT for 95 substrates was possible using standard GP Microplate™, Biolog®, Hayward, CA, USA. The plates were incubated at 28 °C, and colour changes in the tetrazolium dye were followed for 10 days. This allowed for the calculation of a maximum of 95 tolerance values per field community.

4.   **Case 4 (Rutgers et al. 1998b).** The effects of metal exposure on microbial tolerance along the field gradient were determined. Soil samples were taken in March 1997 at five sites with different total metal concentrations. The Biolog® system was applied again.

13.3   PRINCIPLES OF THE FORECASTING METHOD AND LITERATURE TOXICITY DATA

The method used to forecast community-level field effects is the 'added risk approach' (Struijs et al. 1997), essentially based on the SSD approach as described by van Straalen and Denneman (1989).

   In the original method, HCp-values are calculated, which express the ambient concentrations at which a fraction of p% of the species potentially experiences hazard. Benchmark concentrations are often chosen for $p = 5$ (for protection aims) and $p = 50$ (for remediation decisions), and these concentrations are known as HC5 and HC50 (HC = hazardous concentration). The basic methodological assumption of the SSD concept is that sensitivities differ among species, and that the variation in sensitivities in a whole community can be described by a log-logistic (or other) frequency distribution derived from the available toxicity data.

   In practice, (input) toxicity data (NOECs) are obtained in different exposure media. In the current Dutch procedure, they are recalculated into NOEC-values that are assumed to be valid for a standard soil (which is defined to have organic

Table 13.2 Ecotoxicological benchmark concentrations (HC5 and HC50) for zinc in the soils from the case studies, extrapolated using data from single-species laboratory toxicity tests ('species') or breakdown functions ('microbial processes'). The numbers refer to a concentration increase upon the background (given in italic). All values are given in mg kg^{-1}

Metal	Type	HC5	HC50	HC5	HC50	HC5	HC50
		Standard soil		Experimental field plot Zn		Field gradient Budel	
Zinc	Background	(not relevant)		*28*		*11*	
	Species	132	385	57	167	57	166
	Functions	16	207	7	90	7	89

matter and clay contents of 10 and 25% respectively). The formulae are given in VROM (1994). In this way, benchmark concentrations with assumed universal validity are defined. When a HCp-value is applied to a field soil, the inverse formulae are used. The SSD procedure with inclusion of the formulae may be summarized as follows: if a soil has a lower clay and organic matter contents than the standard soil, then the HC5 and HC50 decrease, which implies stricter benchmark concentrations in cases where the sorption capacity is assumed to be low. The formulae are often associated with a scientifically underpinned method for bioavailability correction among soils with different sorption characteristics. That is incorrect, the formulae have no such meaning. Until an underpinned bioavailability correction system is available, the formulae are used in the daily practice of risk assessment, since (numerically) the formulae lead to stricter benchmarks where sorption is considered to be weaker and exposure is expected to be highest.

For the purpose of this chapter, we have had to use existing formulae, since the HCp-values of zinc are given for standard soil only (Crommentuijn *et al.* 1997). The issue of availability is further addressed later.

Frequency distributions are derived separately from toxicity data on 'species' and 'microbial processes'. The notion of 'added' risk is a recent modification of the original method, in which the risk of natural background concentrations is assumed to be negligible. This means that calculated HCp-values refer to generically valid concentration **additions** to the local background, and that true effects at a certain field concentration should be compared to the sum of the HCp-value and the local background.

Benchmark concentrations (HC5 and HC50) were calculated from a recent compilation of toxicity data (Crommentuijn *et al.* 1997), separately for 'species' and 'microbial processes'. Numerical values for the HC5 and HC50 for zinc in standard soil and in the field soils of the case studies are summarized in Table 13.2. Note that it is current practice to quantify benchmark concentrations with total metal concentrations.

In the comparisons between the benchmarks for protection and remediation and the community effects in the case studies, the following notions applied:

• below the HC5, no or minor responses for community endpoints should occur

- exceedance of the HC50 should relate to overt adverse effects for community end points.

Note that these descriptions are still relatively vague. This relates to the fact that risk assessment models usually yield results that can be ecologically meaningful (or not), rather than that they are mechanistically correct (e.g. as in physics). Suter (1993) used the term 'plausibility' of risk assessment models in this context.

13.4   CASE STUDIES

Firstly, a case-by-case evaluation of the data is presented, showing responses of the community end points as well as the underlying changes in species abundance or separate functions (section 13.4). Secondly, the community responses are compared to benchmark concentrations (section 13.5).

13.4.1   CASE 1: NEMATODES IN THE EXPERIMENTAL FIELD PLOT

13.4.1.1   Control performance and overview

The total nematode density in the control enclosures increased from the inoculum density of 600 to maximum enclosure densities of 2875, 5890 and 2198 individuals per 100 g wet soil for the subsequent sampling dates. Overall, 52 taxa were found, representing different trophic groups (fungal feeders, bacterial feeders, plant parasites, and predators (Yeates *et al.* 1993)) and life-history strategies (r–K-like species classification (Bongers 1990)). The maximum number of taxa in a single control replicate was 22 after three months, and this increased to 26 after 22 months. This suggests that species may have colonized the enclosures (e.g. via litter fall), or that densities were initially close to the detection limit.

13.4.1.2   Exposed performance

Patterns of zinc response were grossly similar for different sampling dates; only the data collected after 22 months are shown. The Shannon–Weaver index H was calculated for each enclosure per sampling date. H takes its maximum value when all taxa have equal density, in this case H_{max} would have been 3.26. A value of H near zero indicates overdominance of one species, with rare occurrence of other species. In the field plot, H_{max} was 2.3 (s.e. 0.1) in control conditions. Apparently, species had different densities. Concomitantly with increased total zinc concentrations, H reduced to an average minimum of 0.8 and 1 in the two highest zinc treatments (Figure 13.1, left). At intermediate zinc treatments H tended to increase. This slight increase is apparently related to an increase of taxa number and the increase of overall densities. The increase of H between the two highest treatments is apparently due to a relatively high number of taxa present in the highest concentration, although densities per taxon were, on average, lower (Figure 13.1, right).

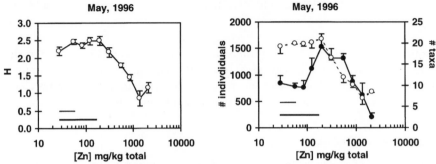

Figure 13.1   Nematode community characteristics in the experimental field plot after 22 months of exposure, expressed by (left) the Shannon–Weaver index H, and by (right) the total density (left y-axis, black symbols) per 100 g soil and the number of taxa (right axis, white symbols). Bars indicate standard errors of five replicates. Benchmark concentrations, extrapolated from laboratory toxicity data, are shown as trajectories (horizontal lines) from the background concentration to the $HC5$ (thin line) or the $HC50$ (thick line). For values, see Table 13.2.

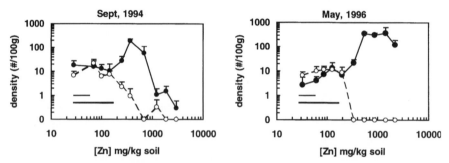

Figure 13.2   The change of density of two nematode taxa in response to zinc exposure in the experimental field plot 3 and 22 months after inoculation. Black markers: *Aphelenchoides* sp.; white markers: *Eucephalobus striatus*. Bars are standard errors of five replicates. When no individuals were encountered in the samples, the marker is set on the x-axis. The horizontal lines are benchmark concentrations (for further information see Figure 13.1).

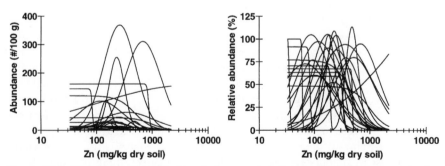

Figure 13.3   The change of the densities of all nematode taxa (smoothed, Y) in response to zinc exposure in the experimental field after 22 months of exposure. Left: true densities per 100 g soil. Right: relative densities (maximum density per species scaled to 100%).

The community endpoints shown in Figure 13.1 integrate the density changes of separate taxa in response to zinc. Figure 13.2 shows the density responses of two separate taxa, to illustrate densities per species, change over time (3 and 22 months of exposure), and variation among replicates. Figure 13.3 shows the smoothed sigmoid or bell-shaped curve fitting results for all separate taxa after 22 months of exposure, to illustrate the density patterns underlying the community changes. The figures show that species indeed have different densities, which change over time (compare Figure 13.2, left and right graph). Responses to zinc ranged from disappearance in all zinc treatments (not visible in Figure 13.3 (left) due to low densities), density reduction at higher zinc treatments and density increases at intermediate concentrations (Figures 13.2, 13.3). For one taxon, *Aphelenchoides* sp., the density increased by more than a factor of 100 compared to the control. Apparently, there are zinc-sensitive and opportunistic species. The sensitive species show a (direct) negative response to zinc exposure, whereas opportunists probably show density increases due to changed ecological interactions (e.g. lack of predation, 'empty' niches). The latter would be an indirect effect of zinc. At high zinc concentrations the opportunists also show a reduced density, probably as a direct effect of these zinc concentrations on population performance.

13.4.1.3 Conclusions

The response of the nematode community to zinc can be quantified using various end points. Detailed analysis shows that zinc caused direct and indirect effects, the latter probably via changed ecological interactions. Moreover, the study shows that integrative end points, such as the Shannon–Weaver index, mask (large) responses of separate taxa.

13.4.2 CASE 2: NEMATODES IN THE FIELD GRADIENT (BUDEL)

13.4.2.1 Reference performance and overview

The most distant sample site in the field gradient was chosen as reference site, and this soil contained 11 mg Zn kg^{-1} dry soil. In this sample, 11 taxa were found, with densities differing between 23 and 397 individuals per 100 g fresh soil (Alkemade *et al.* 1996). Along the sampled gradient, a total of 35 taxa was found.

13.4.2.2 Performance along the gradient

The Shannon–Weaver index H varied between 0.76 and 1.09, while the expected H_{max} was 3.17. As in case 1, taxa had unequal densities. The relationships between zinc exposure and the value of H and the number of taxa or individuals per sample are shown Figure 13.4.

With increasing total zinc concentrations in the soil, it appeared that all three community parameters tended to increase rather than decrease (as expected from Case 1), irrespective of considerable statistical noise. Looking at separate

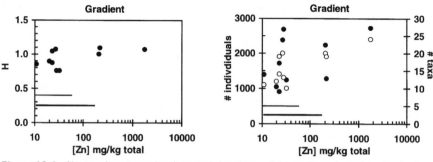

Figure 13.4 Nematode community characteristics along a field gradient of metal contamination, expressed by (left) the Shannon–Weaver index H, and (right) the total densities (left y-axis, black symbols) per 100 g soil and the number of taxa per location (right axis, white symbols). The horizontal lines are benchmark concentrations (for further information, see Figure 13.1).

taxa for which patterns could be discerned, the following trends were observed. Five taxa occurred in all samples, and showed no obvious tendencies of changed densities. Five taxa were found only in the soils with relatively high metal concentrations. Four taxa were absent in the latter soils, but occurred only incidentally in the relatively clean soils. One taxon showed a gross pattern of decreasing density with increasing metal concentrations, and was absent at zinc concentrations higher than approx. 200 mg Zn kg^{-1} dry soil.

How can one interpret these data? Are the observed responses all attributable to zinc, or to the combination of zinc plus other metals that have been emitted in the past? Correlation analyses have shown that zinc was highly correlated with the other metals, but also with soil pH (Posthuma and Notenboom 1996). High metal concentrations were associated with high soil pH. High soil pH may directly cause an increased overall nematode density (see also Figure 13.10). Indirectly, it is likely to reduce the pore water metal concentrations and thereby the bioavailability of metals for pore water dwelling species (Janssen *et al.* 1997b). Both effects counteract the adverse effects of zinc (and other metals) expected from Case 1.

13.4.2.3 Conclusions

The nematode communities along the field gradient showed tendencies for increased H, taxa diversity and total density per sample at higher metal concentrations and higher soil pH. Owing to the association between metals and soil pH, the response cannot be attributed unequivocally to zinc (alone or in combination with other metals).

13.4.3 CASE 3A: MICROBIAL PICT IN THE EXPERIMENTAL FIELD PLOT (^{14}C-ACETATE BREAKDOWN)

13.4.3.1 Reference performance and overview

Forty-eight hours of pre-incubation yielded an approximately equal acetate breakdown activity in all samples without artificial zinc exposure. The relative

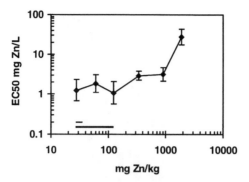

Figure 13.5 Relative tolerance of the microbial communities from the experimental field plot after 19 months, based on acetate mineralization rates as activity measure in the PICT approach. Bars indicate standard errors. The horizontal lines are benchmark concentrations (see further Figure 13.1).

tolerance of the community from the control enclosure was quantified using the artificial exposure concentration that causes 50% activity reduction, i.e. the $EC_{50}^{suspension}$ (mg Zn L^{-1} in incubation conditions), as community end point. For the control soil, the $EC_{50}^{suspension}$ was approximately 1 mg Zn L^{-1} suspension.

13.4.3.2 Exposed performance

The tolerance of the microbial communities from zinc-treated enclosures was established in the same way, and yielded the relationship between community tolerance and the soil zinc concentration in the enclosures (Figure 13.5). Tolerance increased significantly and concomitantly with the concentration of zinc in the soil at concentrations higher than approximately 100 mg Zn kg^{-1} soil.

13.4.4   CASE 3B: MICROBIAL PICT IN THE EXPERIMENTAL FIELD PLOT (BIOLOG® ASSAYS)

13.4.4.1   Control performance and overview

At the standard inoculum density (10^4 colony forming units per well), substrate breakdown activity was typically observed in 80–91 wells. The rate of substrate breakdown activity decreased with increased artificial zinc concentrations in the wells. Tolerance could be established, and was expressed as EC_{50}^{Biolog} (in mg L^{-1} in the wells). The microbial community that was not treated with zinc exhibited highly divergent sensitivities to artificial zinc exposure; the EC_{50}^{Biolog} ranged from 1 to more than 1000 mg Zn L^{-1} for the most and least sensitive breakdown function (Figure 13.6). It is noted that these EC_{50} values should not be compared directly to those of Case 3a, since the zinc availability in the exposure conditions is low, resulting in apparently high EC_{50}^{Biolog} values. In the Biolog® plate, approximately 20% of the added zinc was present in an available form.

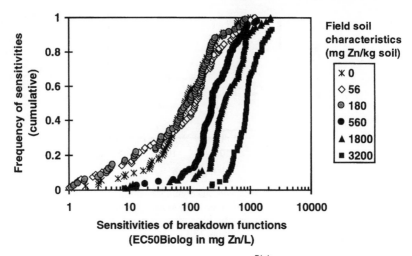

Figure 13.6 Cumulative frequency distributions of EC_{50}^{Biolog} values of separate breakdown functions exhibited by microbial communities from the experimental field plot exposed to different zinc concentrations (indicated right in mg Zn kg^{-1} soil). The HC5 and HC50 derived from microbial processes toxicity data are 7 and 90 mg added Zn kg^{-1}.

13.4.4.2 Exposed performance

Tolerance was calculated for all breakdown functions in the Biolog plate providing information on about 80–91 different community end points (Figure 13.6). At low field zinc concentrations, the tolerance showed only minor changes (note the logarithmic x-axis), which may be attributable to the presence of only small tolerance changes and to the low resolution for EC_{50}-change from the use of a low number of artificial exposure concentrations (six). At higher field concentrations of zinc, however, most functions showed considerably increased tolerance. No function showed a net decrease of tolerance with increasing field zinc concentration, although sometimes a specific function of a specific enclosure demonstrated outlying characteristics within the range of measurement error.

The median EC_{50}^{Biolog}-values of all Biolog substrates of the different field plots are depicted in Figure 13.7 and show that the tolerance increased significantly and concomitantly with the concentration of zinc in the soil at soil concentrations higher than approximately 180 mg Zn kg^{-1} soil.

13.4.4.3 Conclusions Case 3a and 3b

Microbial community tolerance can be quantified using separate breakdown functions, i.e ^{14}C acetate breakdown and the different substrates present in Biolog® plates, but also as the median change of all tolerance values (a maximum of 91) in the latter case. Analyses per function showed that

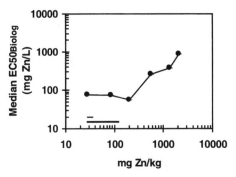

Figure 13.7 Relative tolerance of the microbial communities from the experimental field plot, 33 months after construction. Markers indicate the median EC_{50}^{Biolog} of all functions. Error bars are not given, since the median represents a maximum of 91 different functions with highly different sensitivities (as shown in Figure 13.6).

different functions exhibited different degrees of tolerance evolution. Thus, some community end points (breakdown functions) showed a more sensitive response to zinc exposure than others. In addition, PICT has been established using two different isolation procedures and exposure conditions. Consequently, different parts of the microbial community, analysed by different end points, showed PICT. This contradicts the suspicion that the observed tolerance changes represent the response of a limited part of the total microbial community flourishing in the applied conditions.

13.4.5 CASE 4: MICROBIAL PICT IN THE FIELD GRADIENT (BIOLOG® ASSAYS)

13.4.5.1 Reference performance and overview

In comparison to Case 3b, the number of artificial exposure levels increased from six to eight, and resolution problems for detecting EC_{50}-changes were smaller. Breakdown activity was observed typically in 78–89 wells and tolerance could be quantified for many substrates. For the microbial communities from two relatively clean sites (with 14 and 35 mg Zn kg^{-1} dry soil), the total sensitivity distribution spanned a few orders of magnitude. The EC_{50}^{Biolog} ranged from 0.01 to 2500 mg Zn L^{-1} (Figure 13.8). The upper value was determined by experimental limitations relating to the solubility of zinc in the wells. The median EC_{50}^{Biolog} at these sites was 35 and 16 mg Zn L^{-1} (Figure 13.9), somewhat lower than in Case 3a.

13.4.5.2 Exposed performance

EC_{50}^{Biolog} values were calculated and depicted in cumulative frequency distributions (Figure 13.8). PICT was established. Similar to Case 3b, no tolerance evolution was found for some functions. PICT was established for many functions, no net decreases of tolerance were observed. The median

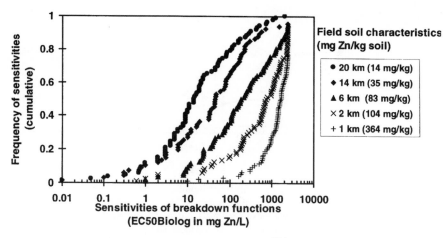

Figure 13.8 Cumulative frequency distributions of EC_{50}^{Biolog} values of microbial breakdown functions in a field gradient of metal contamination. The HC5 and HC50 derived from microbial function toxicity data are 7 and 89 mg added Zn kg^{-1}.

community EC_{50}^{Biolog} increased concomitantly with the field soil zinc concentration, and reached a maximum value for the highest measured soil Zn concentration (364 mg kg^{-1}, Figure 13.9). It should be mentioned that microbial sensitivity for zinc may be influenced by exposure of the microbial communities to other metals present in the gradient, via cross-tolerance evolution (Diaz-Raviña *et al.* 1994). It is, however, expected that co-tolerance played a minor role in the gradient, since all data in this project on single-species tests indicated a dominant role of zinc with respect to toxicity (Posthuma *et al.* 1998b).

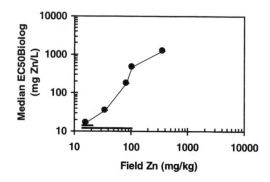

Figure 13.9 Relative tolerance of the microbial communities from the field gradient. Markers indicate the median EC_{50}^{Biolog} of all functions. Error bars are not given, since the median represents a maximum of 89 different functions with highly different sensitivities (see also Figure 13.8).

13.4.5.3 Conclusions

PICT has been established for the microbial communities along the field gradient. In comparison to the nematode data from the same gradient, it is noted that confounding factors have largely been eliminated, which relates (1) to the use of an internal control per sampling site (no zinc added in artificial exposure conditions), and (2) to the use of the compound for which field effects are suspected in artificial exposure experiments. Although confounding factors such as pH may be important (see Case 2), the PICT approach relates zinc exposure in the field to a relative measure of zinc tolerance under standardized laboratory conditions.

13.5 COMPARISONS BETWEEN PREDICTED AND OBSERVED EFFECTS

The comparisons between predicted and observed effects can be made for Cases 1, 3a, 3b and 4. For Case 2, the relationship between metal exposure and effects could not be established unequivocally.

Focusing on the benchmark concentrations in the graphs of the case studies, it can be concluded that toxic effects on overall community parameters are small or negligible at the level of the HC5. At the HC50 level, such effects are either obvious or starting to show a response, exposure is in the latter case near the NOEC of the community end point. Slight increases of the ambient concentration would in such cases induce community responses. At the HC50 level, effects on structural end points were usually less than 50%, as might be expected from the fact that the toxicity data used in the extrapolation procedure concerned NOECs.

From the comparisons, it can be concluded that the HC5 and HC50 predict the potential occurrence of adverse effects on community end points in an ecologically meaningful way. From the data collected in Case 2, it can be concluded that exceeding benchmark concentrations in field conditions is not necessarily associated with observable adverse community effects. In this case, increase of soil pH is likely to have masked the effects of metal exposure on the nematodes. Owing to this confounding soil factor, the data obtained in the nematode gradient study could be used neither to show plausibility nor implausibility of the prediction method.

Looking at separate species or functions, end points may be considerably more sensitive than community end points. This implies that, even when community effects are hardly measurable (community NOEC), toxic effects on sensitive species of functions are present. These responses go unnoticed when looking at community level effects only. This may be attributed to the fact that some species are replaced by others in approximately similar numbers (sensitives versus opportunists), or to the low densities of rare species, which do not contribute much to the numerical change of integrated measures such as the Shannon–Weaver index. These unnoticed changes may have an important ecological meaning. In the nematode studies, for example, it should

Table 13.3 Ecotoxicological benchmark concentrations (HC5 and HC50) for soils, extrapolated using data from single-species laboratory toxicity tests ('species') or breakdown functions ('microbial processes'). Values are given only when relevant for this chapter

Metal	Type	HC5	HC50	HC5	HC50	HC5	HC50
				Field plot Cu/pH		Field gradient Sweden	
Zinc	Background	(not relevant)					n.d.
	Species	132	385			95	278
	Functions	16	207			12	149
Copper	Background			11			n.d.
	Species	24	304	12	157	23	289
	Functions	4	60			3	57

be investigated whether zinc exposure initiates a process of 'immaturation' of the community, i.e. an increase of r-strategists and a concomitant decrease of the Maturity Index (Bongers 1990).

13.6   SCIENTIFIC ASPECTS

13.6.1   LITERATURE DATA ON COMMUNITY EFFECTS IN THE FIELD

The results of the four case studies suggest that the added risk approach yield ecologically meaningful predictions of toxic effects on community endpoints in contaminated soils. This conclusion may be disputed, as it is based upon studies with only zinc. To explore whether the conclusions have a broader validity (other compounds, other soils, other biota), some additional cases from the literature were studied. For these cases, similar procedures have been followed as in our cases. Benchmark concentrations are summarised in Table 13.3.

The first study was carried out in an experimental field with different levels of added copper and different pH (Korthals *et al.* 1996). Nematode communities were studied as in Case 1 and 2, 10 years after artificial contamination. Hazardous concentrations for added copper are shown in Table 13.3. The background concentration of copper was $11 \, \text{mg} \, \text{kg}^{-1}$ soil. As shown in Figure 13.10, pH had a direct influence on nematode densities in soils with low copper concentrations. Considering the effect of copper, nematode densities reduced in the acid soils but not in the soils with a pH of 5.4 and 6.1. This shows that the effects of copper on the nematode numbers depend on copper exposure and pH. Exceeding the added HC50 for copper at pH $\geqslant 5.4$ is not associated with community effects, whereas the same copper concentrations at pH $\leqslant 4.7$ caused considerable effects. In the generic sense, the added HC50 is thus associated with community level effects in worst case conditions. Consequently, the meaning of generic HCs should indeed be qualified as ecologically meaningful predictors of **potential** adverse effects at the two levels of concern. It predicts that some adverse community response is likely to happen in worst-case conditions, but it

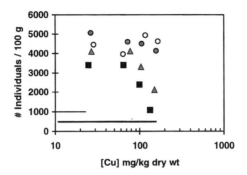

Figure 13.10 Nematode density changes in an experimental field plot with different copper and pH levels, after 10 years of exposure. White markers pH = 6.1; light grey markers pH = 5.4, triangles pH = 4.7, black squares pH = 4.0. The horizontal lines are local added benchmark concentrations for copper (for information, see Figure 13.1).

does not specify **what** will exactly happen and that it will **always** happen. This example suggests further that pH should be considered as an important factor that modulates metal availability and effects.

The second example was executed in a field gradient around a brass works in Sweden (Tyler 1984). Data from various groups of organisms have been compiled, which extends the findings beyond the nematodes and the micro-organisms. Copper and zinc are the major contaminants at the site. The problem of forecasting potential effects for two simultaneously occurring compounds was treated by quantifying the soil metal concentrations as dimensionless contamination units (CU). This approach is similar to the toxic unit approach for studying mixture effects at the species level (Sprague 1970). Concentrations are expressed as dimensionless fractions of a common end point (at the species level often EC_{50}, at the community level both HC5 or HC50 can be used). The data were analysed as in Case study 2. Almost all diversity and activity parameters showed a reduced performance with increasing Cu + Zn concentrations (Figure 13.11). Microfungi and vascular plants showed the weakest responses. At the level of the mixture-HC5, almost no responses of any community end point were visible (data are not shown). At the level of the mixture HC50 for 'species', mosses showed the strongest diversity decline, while other taxa showed negative responses or slight increases at intermediate exposure levels (e.g. plants). Most functional parameters showed a reduction of performance when the mixture HC50 for 'microbial processes' was exceeded.

The analyses of the two literature studies reinforce our conclusion that the 'added risk approach' yields plausible predictions of potential effects. More examples are needed to further confirm this, for example for compounds other than metals. However, as mentioned above, literature data on community responses in field conditions are often scarce or insufficiently precise for further assessments.

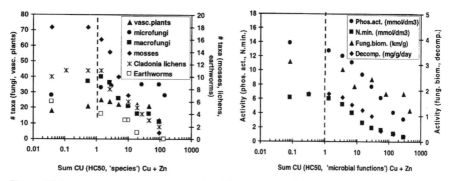

Figure 13.11 Changes of structural (left) and functional (right) community end points of soil biota inhabiting the organic layer of forest soils near a brass mill. Data from Tyler (1984). Effects (Y) are expressed as a function of contamination units (X), by expressing the local soil metal concentrations as dimensionless fractions of their $HC50$. The $HC50$ is given by the dashed vertical line.

13.6.2   TECHNICAL ASPECTS OF THE CASE STUDIES

VARIATION AND STATISTICAL NOISE

In the literature, many data have been filed on the change of species densities or community end points at contaminated field sites (e.g. Bisessar 1982, Bengtsson and Rundgren 1984, Fritze *et al.* 1989, Hågvar and Abrahamsen 1990, Bardgett *et al.* 1994, Koponen and Niemela 1995, Spurgeon and Hopkin 1996, Zaitsev 1998). However, most studies do not cover both the low and high range of exposure, as is needed to assess plausibility of benchmark concentrations like the $HC50$ and the $HC5$. The latter particularly requires accurate data in the low concentration range. Moreover, confounding factors are often present. These induce a considerable statistical noise in the data and disturb straightforward interpretation of response patterns. Plausibility cannot be assessed unequivocally in many cases, as we experienced in Case 2.

In our case studies, designed to compare effects to benchmark concentrations both at a low and a high concentration level, statistical noise and cause–effect problems were minimized as much as possible in the experiment design. Minimizing statistical noise should be a core issue in the design of field studies on community and population parameters. This can be achieved by increasing the number of replicate observations, the number of sampling sites in field gradients, and (for the PICT measurements) the number of artificial treatment levels. Maximum effort in the phase of data collection reduces the chance that random variation disturbs clear data interpretation. In Case 1, for example, the increase of nematode species diversity in the highest zinc treatment of the experimental field plot is most likely to be attributable to chance (immigration) rather than to zinc exposure.

CAUSALITY

The major problem encountered in the interpretation of the case studies was the co-variation between metal concentrations and soil acidity in Case 2.

Consequently, benchmark concentrations for zinc could neither be confirmed nor rejected with these data. Other case studies reported in the literature cannot be properly interpreted for similar reasons. For example, the studies of Zaitsev *et al.* (Zaitsev 1998, Zaitsev *et al.* 1998) near a metallurgic plant showed changes in the community structure of oribatid mites. Just as in Case 2, these responses could have been caused by metals as well as soil pH. An unclear cause–effect relation in case studies can be considered as one of the major scientific reasons for not using field data in deriving ecotoxicological benchmark concentrations. The PICT methodology presents an elegant solution to these interpretation problems. Firstly, because the internal control of the artificial exposure procedure balances between-sites variability that normally occurs. Secondly, because the use of the supposedly toxic compound in the artificial exposure studies strengthens the link between effect and assumed cause.

BIOAVAILABILITY

In the case studies described in this chapter, bioavailability has not been taken into consideration in a mechanistically meaningful way. It is well known that total soil concentrations, as used in this chapter, are not directly associated with uptake or toxic effects in organisms (Allen 1997, Posthuma *et al.* 1998a). For many species, especially the soft-bodied ones, the concentrations in the soil pore water are important. These concentrations depend on total concentrations and are modulated by soil characteristics, the most important one being pH (Janssen *et al.* 1997a). At low pH, metal sorption is low and exposure of pore water exposed organisms is high. Ageing generally reduces metal availability by increased sorption to the mineral lattice. With this in mind, the dissimilarities in the responses between the case studies are problematic to explain for the nematodes (1 and 2) because of the dual effect of soil acidity on both bioavailability and species performances in the gradient. On the other hand, the microbial tolerance data (Cases 3 and 4) clearly suggest that microbial responses were largest in the field gradient. This may have been caused by a relatively high pore water concentration in the gradient soils with a low pH, but also by the extended exposure period in the gradient.

STRUCTURE AND FUNCTION

In relation to single species toxicity tests, diversity parameters (such as those used in this chapter) are classical and ecologically meaningful community end points to quantify community-level structural changes in the field. However, for assessing functional changes in microbial communities, PICT is a new technique for soil organisms. We consider PICT an ecologically relevant community end point. Doelman *et al.* (1994), for example, showed that the versatility of breakdown functions exhibited by tolerant microbial communities reduces in the process of gaining tolerance. Our data, moreover, suggest that the frequency distribution of sensitivities is narrowed in tolerant communities. It is as yet unclear whether this implies a decreased diversity of microbial species executing these functions. For an estuarine nematode community in which both species diversity and community tolerance were established, PICT coincided with decreased species diversity (Millward and Grant 1995). Further work should substantiate the ecological meaning of increased PICT in microbial communities.

13.6.3 CONCLUSIONS ON SCIENTIFIC ASPECTS

The design of the case studies did not pose problems to quantify community response parameters per se, or the underlying responses of separate taxa or microbial functions. Each end point, be it community or population level, showed its own characteristic response that ranged from highly sensitive to insensitive and even opportunistic. The latter species showed an increased density at high exposure concentrations. For the nematodes, community diversity end points were less sensitive than the underlying density changes of various species. For the micro-organisms, large sensitivity differences were present among the different functions. The causal chain, however, was not clear in the case study on the nematode community structure in the field gradient. The case study data have been worked out in more detail elsewhere (Posthuma *et al.* 1998b), and can be further interpreted in view of, for example, the role of pH in metal bioavailability, direct and indirect effects, the role of genetic adaptation of the nematodes, etc.. Within the framework of the existing risk assessment methodology we have only used summary data related to total concentrations.

13.7 POLICY

13.7.1 REMAINING UNCERTAINTIES AND ASSUMPTIONS

For policy use, it is important to know the general validity of the conclusion that the methodology yields ecologically meaningful benchmark concentrations, for all soils, compounds and soil ecosystems. The question remains as to whether the concomitance between observed and forecasted effects is coincidental or systematic. A somewhat speculative answer might be found in a hidden aspect of the forecasting method, related to the meaning of the term 'frequency distribution of sensitivities'. Sensitivity differences among species (or functions) are handled in the statistical extrapolation step. Often, the laboratory toxicity data are considered to represent 'species sensitivities' per se. However, toxicity tests are executed in various substrates with different sorption characteristics, with recently added compounds. Consequently, the calculated sensitivity distribution represents the 'sensitivities of species including substrate variability'. Further, the environments within the test often represent worst-case conditions considering that the compounds are freshly added. This may introduce a worst-case bias in the sensitivity distribution. For example, the sensitivity distribution for copper has been established with various species in various soils, and it does not precisely predict the toxic effects on nematodes at neutral soil pH (Figure 13.10). However, it does predict **potential** adverse effects on community end points, which would become overt if soil conditions change to worst case. In this sense, the generic forecasting method relates to a particular policy aim, namely the governmental desire to avoid a false-negative label to a soil with effective hazardous toxicant concentrations (Denneman 1998).

It has already been noted, that current benchmark concentrations are based on imperfect toxicity data. Despite the plausibility of benchmark concentrations in forecasting potential effects, there is a general wish to improve the forecasting quality, both scientifically (uncertainties should be reduced whenever possible), and governmentally (scientific underpinning of benchmark concentrations used to motivate expensive cleanup decisions). To reduce uncertainties in effect forecasting, various options are available:

1. Uncertainties in the toxicity data on the compound under investigation should be minimized. Accurate bioavailability correction will considerably improve the field relevance of toxicity data used in the forecasting procedure and of the benchmark concentrations themselves. In the laboratory toxicity data of the compounds in our case studies, pH was, for example, neglected as an important factor influencing metal availability. Provisional calculations yielded similar conclusions on the plausibility of generic benchmark concentrations, with and without taking clay and organic matter contents into account (data not shown, but the changes of the benchmarks relative to the response curves were small for the Figures 13.1, 13.2, 13.4, 13.5, 13.7, 13.9–13.11). However, correction for pH differences among field soils is likely to be important, and would improve predictive accuracy in site-specific assessments (Janssen *et al.* 1997a).

2. Implicit assumptions of the statistical extrapolation procedure should be checked for each compound. Lack of fit may occur, for instance a bimodal sensitivity distribution for specific working compounds such as pesticides. Data analyses for the metals (Crommentuijn *et al.* 1997) have shown that there were no significant deviations from a log-logistic sensitivity distribution. Other mathematical approximations than the log-logistic distribution may also be applied, since there are no ecological reasons for choosing one or the other model.

3. Uncertainty may be reduced by considering additional groups of organisms, with particular interest for groups that require special attention, for example because they occur at contaminated sites, or they have special protection status. For the purpose of evaluating generic benchmark concentrations, our case studies focused on nematodes and micro-organisms. Gross comparisons did not reveal an obvious hyper- or hyposensitivity of these groups compared to the laboratory-tested 'average' species from which the sensitivity distribution is constructed.

13.7.2   CONCLUSIONS ON POLICY

Effect forecasts generated with the added risk approach appear to be plausible predictors of potential adverse effects on biotic communities at contaminated sites, with a limited chance of yielding false negatives. The forecasting system can (and should only) be used for the first-tier assessment of risks, considering core questions of the (limited) type: 'is it likely that we have an environmental problem here?'. Second-tier assessment methods should be developed in order to specify the risks at particular contaminated sites, whereby site-specific information should be used to assess the chance for adverse effects given the local conditions and species groups occurring. Overall, we concluded that uncertainties in effect forecasting can certainly be reduced (Posthuma *et al.* 1998b), starting in the phase of data collection, but simultaneously it is clear that the first-tier risk assessment methodology is sufficient for its (restricted) aims. After a fast discrimination between 'unlikely to be affected' ($<$HC5) and 'potentially seriously affected' ($>$HC50) in the first tier, second tier assessments can be executed, focusing on the latter sites only. In the second tier assessments, the important local ecosystem variables can be taken into account, to indicate whether the potential risks already noticed have induced true adverse effects in the local ecosystem, e.g. due to low soil pH or due to the local occurrence of sensitive species.

ACKNOWLEDGEMENTS

We thank the convenors of the SETAC-UK conference on 'forecasting' for their invitation to prepare this chapter. Special thanks are due to all people involved in the research project, Validation of Toxicity Data and Benchmark Concentrations for Soils, which was executed as a collaboration between RIVM (the Dutch National Institute of Public Health and the Environment), the Vrije Universiteit Amsterdam (Faculty of Biology, Department of Ecology and Ecotoxicology) and TNO (the Netherlands Organization for Applied Research) at the request of the Dutch Ministry of Housing, Spatial Planning and the Environment. In this respect, special thanks are due to those involved in facilitating or executing the community-end point work, namely Rob Alkemade, Rob Baerselman, Ton Breure, Bea Wind, Herman Eijsackers, Arja Fleuren-Kemilä, Arie-Jan Folkerts, Gerard Korthals, Jos Notenboom, Olivier Klepper, Tycho Schol, Els Smit, Baukje Sweegers, Mariette Van Esbroek, Kees Van Gestel, Ilse Van 't Verlaat, Rens Van Veen, Kees Verhoef and Hans Vonk.

REFERENCES

Alkemade R, Schouten AJ, Kersten P and van Esbroek MLP (1996) *Vergelijking van Effectniveaus voor Bodemorganismen met het Maximaal Toelaatbaar Risico. Een Veldstudie naar de Invloed van Zware Metalen in een Gradient te Budel*. National Institute of Public Health and the Environment, Bilthoven, the Netherlands, RIVM, Report No. 607505001 (in Dutch).
Allen HE (1997) Standards for metals should not be based on total concentrations. *SETAC-Europe News*, **8**, 7–9.
Bardgett RD, Speir TW, Ross DJ, Yeates GW and Kettles HA (1994) Impact of pasture contamination by copper, chromium, and arsenic timber preservative on soil microbial properties and nematodes. *Biology and Fertility of Soils*, **18**, 71–79.
Bengtsson G and Rundgren S (1984) Ground-living invertebrates in metal-polluted forest soils. *Ambio*, **13**, 29–33.
Bisessar S (1982) Effect of heavy metals on microorganism in soils near a secondary lead smelter. *Water, Air and Soil Pollution*, **17**, 305–308.

Blanck H, Wängberg S-Å and Molander S (1988) Pollution-induced community tolerance — a new ecotoxicological tool. In *Functional Testing of Aquatic Biota for Estimating Hazards of Chemicals*, Cairns J Jr and Pratt JR (eds), ASTM STP 988, American Society for Testing and Materials, Philadelphia, pp. 219–230.

Bongers T (1990) The maturity index: an ecological measure of environmental disturbance based on nematode species composition. *Oecologia*, **83**, 14–19.

Chapman PM (1995a) Extrapolating laboratory toxicity results to the field. *Environmental Toxicology and Chemistry*, **14**, 927–930.

Chapman PM (1995b) Do sediment toxicity tests require field validation? *Environmental Toxicology and Chemistry*, **14**, 1451–1453.

Crommentuijn T, Polder MD and van De Plassche EJ (1997) *Maximum Permissible Concentrations and Negligible Concentrations for Metals, taking Background Concentrations into account*. National Institute of Public Health and the Environment, Bilthoven, the Netherlands, RIVM, Report No. 601501001.

Denneman CAJ (1998) Aanpassing van bodemnormen? Biobeschikbaarheid als onderdeel van een norm of als flexibel element in de risicobeoordeling. *Bodem*, **11**, 23–25 (in Dutch).

Diaz-Raviña M, Bååth E and Frostegård Å (1994) Multiple heavy metal tolerance of soil bacterial communities and its measurement by a thymidine incorporation technique. *Applied Environmental Microbiology*, **60**, 2238–2247.

Doelman P, Jansen E, Michels M and van Til M (1994) Effects of heavy metals in soils on microbial diversity and activity as shown by the sensitivity-resistance index, an ecologically relevant parameter. *Biology and Fertility of Soils*, **17**, 177–184.

Fritze H, Nini S, Mikkola K and Makinen A (1989) Soil microbial effects of a copper nickel smelter in southwestern Finland. *Biology and Fertility of Soils*, **8**, 87–94.

Hågvar S and Abrahamsen G (1990) Microarthropoda and Enchytraeidae (Oligochaeta) in naturally lead-contaminated soil: A gradient study. *Environmental Entomology*, **19**, 1263–1277.

Hopkin SP (1993) Ecological implications of '95% protection levels' for metals in soil. *Oikos*, **66**, 137–141.

Janssen RPT, Peijnenburg WJGM, Posthuma L and van Den Hoop MAGT (1997a) Equilibrium partitioning of heavy metals in Dutch field soils. I. Relationships between metal partition coefficients and soil characteristics. *Environmental Toxicology and Chemistry*, **16**, 2470–2478.

Janssen RPT, Posthuma L, Baerselman R, Den Hollander HA, van Veen RPM and Peijnenburg WJGM (1997b) Equilibrium partitioning of heavy metals in Dutch field soils. II. Prediction of metal accumulation in earthworms. *Environmental Toxicology and Chemistry*, **16**, 2479–2488.

Kooijman SALM (1987) A safety factor for LC50 values allowing for differences in sensitivity among species. *Water Research*, **21**, 269–276.

Koponen S and Niemela P (1995) Ground-living arthropods along pollution gradient in boreal pine forest. *Entomologica Fennica*, **6**, 127–131.

Korthals GW, Alexiev AD, Lexmond ThM, Kammenga JE and Bongers T (1996) Long-term effects of copper and pH on the terrestrial nematode community in an agroecosystem. *Environmental Toxicology and Chemistry*, **15**, 979–985.

Lande R (1996) Statistics and partitioning of species diversity, and similarity among multiple communities. *Oikos*, **76**, 5–13.

Millward RN and Grant A (1995) Assessing the impact of copper on nematode communities from a chronically metal-enriched estuary using pollution-induced community tolerance. *Marine Pollution Bulletin*, **30**, 701–706.

Posthuma L (1992) Genetic ecology of metal tolerance in Collembola. PhD thesis, Vrije Universiteit Amsterdam, the Netherlands.

Posthuma L (1997) Effects of toxicants on population and community parameters in field conditions, and their potential use in the validation of risk assessment methods. In *Ecological Risk Assessment of Contaminants in Soil*, van Straalen NM and Løkke H (eds), Chapman and Hall, London, pp. 85–123.

Posthuma L and Notenboom J (1996) Toxic effects of heavy metals in three worm species (*Eisenia andrei, Enchytraeus crypticus* and *Enchytraeus albidus*: Oligochaeta) exposed in artifcially polluted soil substrates and contaminated field soils. National Institute of Public Health and the Environment, Bilthoven, the Netherlands, RIVM, Report No. 719102048.

Posthuma L, Traas TP and Suter GW (eds, in prep.) *The Use of Species Sensitivity Distributions in Ecotoxicology*. SETAC Press, Pensacola, FL, USA.

Posthuma L, Notenboom J, De Groot AC and Peijnenburg WJGM (1998a) Soil acidity as major determinant of zinc partitioning and zinc uptake in two oligochaete worms (*Eisenia andrei* and *Enchytraeus crypticus*) exposed in contaminated field soils. In *Progress in Earthworm Ecotoxicology*, Sheppard SC, Bembridge JD, Holmstrup M and Posthuma L (eds), SETAC Press, Pensacola, FL, USA, pp. 111–127.

Posthuma L, Van Gestel CAM, Smit CE, Bakker DJ and Vonk JW (1998b) *Validation of Toxicity Data and Risk Limits for Soils: Final Report*. National Institute of Public Health and the Environment, Bilthoven, the Netherlands, RIVM, Report No. 607505004.

Rutgers M, Seegers BMC, Wind BS, van Veen RPM, Folkerts AJ, Posthuma L and Breure AM (1998a). Pollution-induced community tolerance in terrestrial microbial communities. In *Contaminated Soil. Proc. 6th Int. FZK/TNO Conference on Contaminated Soil, May 17–21, 1998 — Edinburgh*, London, Thomas Telford Publishing, pp. 337–343.

Rutgers M, Van 't Verlaat IM, Wind B, Posthuma L and Breure AM (1998b) Rapid method to assess pollution-induced community tolerance in contaminated soil. *Environmental Toxicology and Chemistry*, **17**, 2210–2213.

Schouten AJ, van Esbroek MLP and Alkemade JRM (1998) Effects of zinc on communities and biodiversity of soil nematodes. In *Validation of Toxicity Data and Risk Limits for Soils: Final Report*, Posthuma L, van Gestel CAM, Smit CE, Bakker DJ and Vonk JW (eds), National Institute of Public Health and the Environment, Bilthoven, the Netherlands, RIVM, Report No. 607505004, pp. 131–146.

Smit CE, van Beelen P and Van Gestel CAM (1997) Development of zinc bioavailability and toxicity for the springtail *Folsomia candida* in an experimentally contaminated field plot. *Environmental Pollution*, **98**, 73–80.

Smit CE and van Gestel CAM (1996) Comparison of the toxicity of zinc for the springtail *Folsomia candida* in artificially contaminated and polluted field soils. *Applied Soil Ecology*, **3**, 127–136.

Sprague JB (1970) Measurement of pollutant toxicity to fish. II. Utilizing and applying bioassay results. *Water Research*, **4**, 3–32.

Spurgeon DJ and Hopkin SP (1996) The effects of metal contamination on earthworm populations around a smelting works: quantifying species effects. *Applied Soil Ecology*, **4**, 147–160.

Struijs J, van De Meent D, Peijnenburg WJGM, van Den Hoop MAGT and Crommentuijn T (1997) Added risk approach to derive maximum permissible concentrations for heavy metals: how to take into account the natural background levels? *Ecotoxicology and Environmental Safety*, **37**, 112–118.

Suter GW (1993) *Ecological Risk Assessment*. Lewis, Boca Raton, FL, USA.

Tyler G (1984) The impact of heavy metal pollution on forests: a case study of Gusum, Sweden. *Ambio*, **13**, 18–24.

USEPA (1978) United States Environmental Protection Agency. Fed. Regist. 43:21506-21518. May 18.

van Beelen P, Vonk JW and Rutgers M (1998) Effects of zinc on soil microbial mineralization and microbial communities. In *Validation of Toxicity Data and Risk Limits for Soils: Final Report*, Posthuma L, van Gestel CAM, Smit CE, Bakker DJ and Vonk JW (eds), National Institute of Public Health and the Environment, Bilthoven, the Netherlands, RIVM, Report No. 607505004, pp. 147–161.

van Straalen NM (1993) An ecotoxicologist in politics. *Oikos*, **66**, 142–143.

van Straalen NM and Denneman CAJ (1989) Ecotoxicological evaluation of soil quality criteria. *Ecotoxicology and Environmental Safety*, **18**, 241–251.

VROM (1994) *Environmental Quality Objectives in the Netherlands*. Dutch Ministry of Housing, Spatial Planning and the Environment, Samsom-Sijthoff Press, Alphen a/d Rijn, the Netherlands.

Yeates GW, Bongers T, De Goede RGM, Freckman DW and Georgieva SS (1993) Feeding habits in nematode families and genera — an outline for soil ecologists. *Journal of Nematology*, **25**, 315–332.

Zaitsev AS (1998) Populations of oribatid mites in the surroundings of Kosaya Gora near a metallurgic plant. In *Pollution-induced Changes in Soil Invertebrate Food-webs*, Butovsky RO and Van Straalen NM (eds), Dept. Ecology and Ecotoxicology, Vrije Universiteit, Amsterdam, the Netherlands, pp. 45–53.

Zaitsev AS, Verhoef SC, Pokarzevskii AD, Filimonova ZV and Butovsky RO (1998) General description of the Kosaja Gora research area. In *Pollution-induced Changes in Soil Invertebrate Food-webs*, Butovsky RO and Van Straalen NM (eds), Dept. Ecology and Ecotoxicology Vrije Universiteit, Amsterdam, the Netherlands, pp. 31–44.

14

Prediction of Sediment Quality in the Dutch Coastal Zone: Model Validation and Uncertainty Analysis for Cd, Cu, Pb, Zn, PCBs and PAHs

H. L. A. SONNEVELDT AND R. W. P. M. LAANE

National Institute for Coastal and Marine Management (RIKZ), The Hague, The Netherlands

14.1 INTRODUCTION

The concentrations of metals and organic micropollutants in sediments of the Dutch coastal zone are among the highest found in the entire North Sea (Salomons *et al.* 1988). As a result, ecological effects have been apparent, as in the cases of fish diseases (Vethaak 1993) and effects on the reproduction of marine mammals in the Wadden Sea (Reijnders 1996). The potential risk of negative ecological effects is still present (Tijink 1998).

Recently, decreasing trends in the particulate concentrations of metals, polychlorinated biphenyls (PCBs) and (less so) for PAHs have been found in the surface sediments of the Dutch coastal zone (Laane *et al.* 1999). For the metals, the reduction percentages between 1981 and 1996 varied between 71% for Cd, 28% for Cu, 49% for Zn and 53% for Pb. For PCBs and PAHs the reduction percentages between 1986 and 1996 were 69% and 26% respectively. To link the concentration in the surface sediments with emission data, a model for Cd and PCBs has been developed by Laane *et al.* (1999). The value of such a model in forecasting potential ecotoxicological effects is clear. It can be used to predict changes in the concentrations of metals and organic compounds in the surface sediments in terms of changing loads from different sources, for example the dumping of dredged materials.

The model assumes that the main process causing the cleaning up of the surface sediments is the physical exchange of old mud particles with higher concentrations of compounds in the surface sediment with fresh mud particles with lower concentrations of compounds suspended in the water column. The Dutch coastal zone is a dynamic environment (Eisma 1988). No net sedimentation occurs, with the exception of the harbours and estuaries. The median contribution of the $<63\,\mu$m fraction in the sandy sediments of the

Forecasting the Environmental Fate and Effects of Chemicals. Edited by Philip S. Rainbow, Steve Hopkin and Mark Crane.
© 2001 John Wiley & Sons Ltd

coastal zone is approximately 2% (Laane *et al.* 1999). The <63 μm particle fraction is defined as mud (Alphen 1990). The residence time of mud particles with associated compounds in the sediment is determined by the mass of mud particles in the active bottom layer and the magnitude of the mud balance for the coastal zone. The mass of mud particles in the sediment was estimated as 29 Mt; this follows from the thickness of the active layer (40 cm), the mud percentage (2%), the dry density (1400 kg m^{-3}) and the surface area of the model box (Laane *et al.* 1999). The annual flux of mud through the water column of coastal zone was established as 10 Mt, which consists of 6.5 Mt from southern input and 3.5 Mt from river inputs and the dumping of dredged materials (Alphen 1990). The half-life of mud particles and associated compounds in the active bottom layer follows from these figures and equals about two years.

The model was able to simulate the observed particulate concentrations of Cd and PCBs in the sediments of the Dutch coastal zone for the period 1980–94 (Cd) or 1986–94 (PCBs) rather well (Laane *et al.* 1999). It is remarkable that this model does not include any (bio-)chemical reactions, since it is assumed that the particulate fraction behaves conservatively. It is noted that desorption and volatilization, especially of lower chlorinated PCBs and lower molecular weight PAHs, are important processes in several water systems, for example an English lake (Gevao *et al.* 1997), Chesapeake Bay (Nelson *et al.* 1998), the Hudson river estuary (Achman *et al.* 1996) and the Great Lakes (Hillery *et al.* 1998). PCBs in sediments may also be subject to biodegradation (e.g. Brown *et al.* 1987, Brown and Wagner 1990). However, mobilization processes lead to an enrichment of higher chlorinated PCBs and higher molecular weight PAHs in the sediment, which has not been observed in the surface sediments of the Dutch coastal zone (Laane *et al.* 1999). This suggests that (chemical) remobilization does not play a major role in the fate of PCBs and PAHs in the surface sediments of the Dutch coastal zone.

The predictive value of a model can be evaluated by checking its ability to reproduce (independent) historical observations (validation), and by testing the sensitivity of the model to its uncertainties. The model was validated for Cd and PCBs by comparing the simulated sediment concentrations with observed concentrations. The field data are an independent data set (they are not used to determine model input data). Main sources of uncertainty in water quality models are the model formulations (parameters and processes), and the loads of the compounds. The model formulations may be incomplete (loss processes lacking) or incorrect, or the parameterization may be incorrect (wrong mixing depth). Estimates of annual average loads of compounds into the North Sea are uncertain since data are scarce and often inaccurate.

In this chapter the validation of the model is extended to other compounds. Given the availability of data, the following compounds have been selected besides Cd and PCBs: Cu, Pb, Zn and PAHs. Furthermore the sensitivity of the model for uncertainty in the emission data will be tested, and the sensitivity of the model for the model formulations and model parameters will also be investigated. Finally, the model is applied to evaluate the effect of a reduction in the dumping of dredged materials. Any such further validation of the model will provide more confidence in its use in environmental forecasting.

14.2 THE DUTCH COASTAL ZONE AS A CASE STUDY

The analysis is restricted to the first 20 km seawards of the Dutch coastal zone, north of the Rhine outflow (Figure 14.1). The length of this box is 130 km, the surface area 2600 km². In this area the river Rhine mixes with North Sea water, and the fresh water plume remains confined to a relatively narrow area along the coastal zone (Ruijter et al. 1992). The general direction of the residual current is from the south to the north.

The compound concentrations in the surface sediments of the Dutch coastal zone were analysed in the smaller than 63 µm fraction; sediment samples were taken from the upper 20 cm with a grab (Laane et al. 1999). Field data were available for 1981, 1986, 1991 and 1996 for metals and 1986, 1991 and 1996 for PCBs and PAHs. PCBs is the sum of 7 PCBs: 52, 101, 118, 138, 153, 170 and 180 (Laane et al. 1999). PAHs is the sum of 6 PAHs: fluoranthene, benzo(b)fluoranthene, benzo(k)fluoranthene, benzo(a)pyrene, indeno(1,2,3-cd)pyrene and benzo(ghi)perylene (Laane et al. 1999). The range in the field data is expressed as the highest value, being the median plus the standard deviation and the lowest value, being the median minus the standard deviation.

The choice of the simulation period was based on the availability of load data. For metals the period was 1980–94, for PCBs 1986–94 and for PAHs 1988–94. The starting concentration of the compounds in the surface sediment was the total load from the different sources divided by the annual flux of suspended matter through the coastal zone.

14.2.1 PARTICULATE LOADS TO THE DUTCH COASTAL ZONE

The following sources of particulate metals and particulate organic compounds to the Dutch coastal zone were distinguished: southern input (mainly Atlantic Ocean via the Channel), rivers (with estuarine retention taken into account), dumping of dredged materials from harbours (mainly the port of Rotterdam) and atmospheric deposition. The main contribution to the source 'rivers' is from the Rhine–Meuse estuary; it also includes the Scheldt estuary and other freshwater sources such as the discharge sluices at IJmuiden (Figure 14.1).

The annual average southern input is assumed to be constant over the years. The load of the compounds studied is calculated from the average annual flux of suspended matter from the south to the Dutch coastal zone of 6.5 Mt yr^{-1} (Alphen 1990), and average concentrations of metals and organic compounds in suspended matter in the Straits of Dover, which are taken from James et al. (1993) and Ras et al. (1996) respectively.

The annual average atmospheric deposition was quantified with data on direct measurement of dry deposition (Otten 1991); the trend in time was based on data

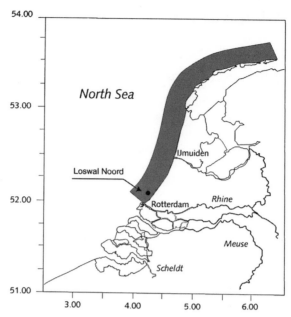

Figure 14.1 The Dutch coastal zone with the model box (shaded area). The main dumping location ('Loswal Noord') for dredged materials is located near the outflow of the river Rhine.

from Hoornaert *et al.* (1996). The annual average particulate load of compounds to the North Sea from the freshwater sources (rivers and discharge sluices) in the Netherlands is calculated from particulate concentrations and water fluxes from Dutch monitoring locations in the freshwater part of the estuary (Anonymous 1997). These river loads were corrected for estuarine retention (Wulffraat *et al.* 1993, Zwolsman 1994).

The annual average load of metals and organic compounds to the Dutch coastal zone by the dumping of dredged materials from harbours for the different years is taken from Lourens (1996). The main contribution is dredged materials from Rotterdam harbour, which are dumped at the 'Loswal Noord' dumping site (Figure 14.1). The load data were corrected for the retention of particulate matter at this site by burial and by taking into account the return flow of mud towards Rotterdam Harbour. The retention was about 70% before 1985 and some 50% from 1985 onwards (Spanhof *et al.* 1990). In the model, it is assumed that during the dumping of dredged materials in the coastal zone the compounds behave conservatively (Hegeman and van der Weijden 1990), so no adsorption or desorption of the compounds on suspended matter is taken into account.

14.2.2 UNCERTAINTY IN LOAD DATA

For a mass balance of dredged material with associated compounds of the Rotterdam harbour area uncertainties of the loads have been estimated already:

Rhine outflow: Cd±35%, Cu, Zn, Pb and PAHs±48 % (Tijink 1995); total mass balance for Cd±46% (Tijink *et al.* 1993) and for PCBs at least ±60% (Wulffraat *et al.* 1993). Following these previous estimates, we assume that the uncertainty in the annual average load to the Dutch coastal zone will be at least ±50 % for heavy metals and ±60 % for PCBs and PAHs.

The uncertainty will be different for the different sources, because of the nature of the source and the availability of data. For rivers and dumping of dredged materials relatively many data are available, compared to the southern input and atmospheric deposition.

The discharge and contaminant concentrations of the main river outflows are monitored intensively (two-week intervals). Still, the calculated loads contain a considerable amount of uncertainty (Vries and Klavers 1994). The precision of the calculated annual average load of suspended matter in the river Rhine at the Dutch–German border is 30% at a two-week sampling intervals (Vries and Klavers 1994). This implies that the overall uncertainty in the calculated annual average load of substances in suspended matter would be at least 30% here. Moreover, a significant fraction of the Rhine outflow follows the discharge sluices in the Haringvliet barrier dam. The quantification of loads through sluice systems is not yet well established (Vries and Klavers 1994).

Uncertainties in, for example, partitioning coefficients and net sedimentation rates, cause uncertainty in the calculated filtering capacities of the estuaries. However, the magnitude of the uncertainty could not be established (Zwolsman 1994).

The uncertainty in atmospheric deposition is considerably larger. For emission-based model calculations of the yearly averaged atmospheric deposition, the uncertainty is estimated as factors of 1.5–2.5 for metals, 2–5 for PCBs and 2–3 for PAHs (Baart *et al.* 1995).

Random errors and spatial and temporal variability in measured concentrations and amounts dredged cause uncertainty in the total load of compounds through dumping of dredged materials, the relative error for Cd is estimated as 42% (Tijink *et al.* 1993). The error in the correction factor for retention at the dump site and return flow is not known.

The uncertainty in the annual average loads of water, suspended matter, metals and organic compounds through the English Channel will be considerable, since water flux and concentrations are not monitored systematically (Laane *et al.* 1996). For this reason, it is to be expected that the uncertainty in the southern input will be comparable to the uncertainty in atmospheric deposition.

14.2.3 PARTICULATE LOADS

The calculated annual average particulate loads of Cd, Cu, Pb, Zn, PCBs and PAHs to the Dutch coastal zone through southern input, rivers, atmosphere and dumping of dredged materials are shown in Figure 14.2. The period covered is 1980–94 for metals, 1986–94 for PCBs, and 1988–94 for PAHs. The annual

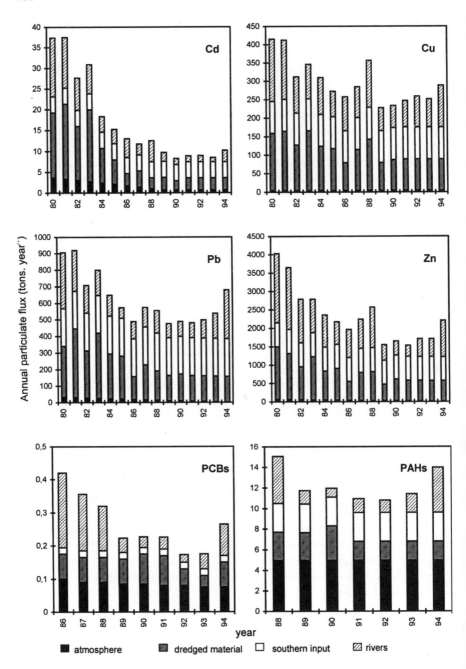

Figure 14.2 Annual average particulate load (t yr^{-1}) to the Dutch coastal zone of Cd, Cu, Pb and Zn (1980–94), PCBs (1986–94) and PAHs (1988–94). Sources are atmosphere, dumping of dredged material, southern input and rivers.

average particulate loads of the metals decrease between 1980 and 1986 (Cu, Pb and Zn) or 1990 (Cd), and stabilize or even slightly increase afterwards. The decrease is about 70% for Cd, 35% for Cu, 40% for Pb and 50% for Zn over the studied period. The particulate annual average load of PCBs drops by approximately 50% between 1986 and 1989 and then stabilizes. The particulate annual average load of PAHs does not show a clear trend between 1988 and 1994.

The relative contribution of the annual average loads from the different sources changes over the period studied, and varies for the different compounds. For the metals, the summarized contributions of dredging and rivers to the total particulate load decreased between 1980 and the 1990s, from 80% to 50% for Cd and Zn, from 75% to 65% for Cu, and from 70% to 60% for Pb. For PCBs, this contribution decreased from about 70% (1986) to about 50% (1990s). For PAHs, dredging and rivers account for 30–50% of the total particulate load.

The annual average load from southern input becomes relatively more important for the metals during the period studied. For instance, for Cd the contribution to the total annual average particulate load increases from about 10% in 1980 to about 40% in 1994. For PCBs the southern input is relatively unimportant (5–8%); for PAHs the contribution is approximately 20% during the period studied. Atmospheric deposition seems to be a minor source for particulate metals in the Dutch coastal zone; the largest contribution is about 10% for Cd. The relative contribution of atmospheric deposition is much more important for PCBs (about 25%) and PAHs (about 40%).

14.3 MODEL SIMULATIONS

The simulated concentration, the range in observed concentrations, and the annual average particulate loads of Cd, Cu, Pb, Zn, PCBs and PAHs are shown in Figure 14.3 for the different years. For all compounds, the model results agree rather well with the observed values. The delay between the annual average load reduction and the simulated concentration is about two and a half years.

For Cd, Cu and Zn, the fit between simulated and observed concentrations is remarkably good. The concentration level as well as the trend are reflected by the model. For Pb, the observed trend is simulated correctly, but the concentration level seems to be underestimated by about 20%. The PCBs simulation also predicts the observed trend, but here the level is overestimated by 30–40%. The observed PAHs concentrations show a slight decrease, although the order of magnitude is predicted correctly. The model overestimates the concentrations by a factor of 2.

The question remains how sensitive the model is to the uncertainties which have been discussed above. This was tested for a metal (Pb) and a category of organic compounds (PCBs). The influence of uncertainty in the model formulations was investigated by variation with 10 cm of the mixing depth (40 cm) for Pb and by incorporating a first-order loss process for PCBs with a rate

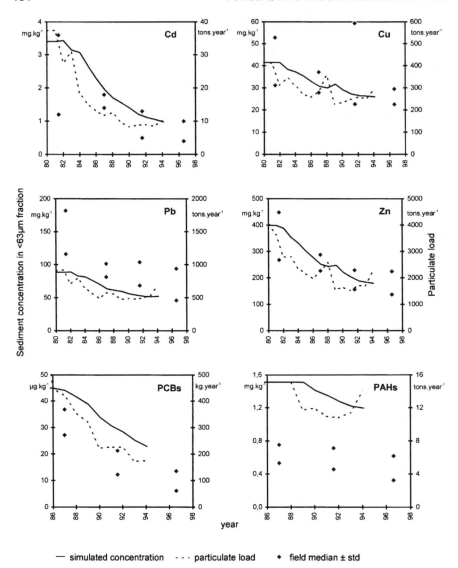

Figure 14.3 Simulated and range of observed particulate concentrations in the Dutch coastal zone (left vertical axis; mg kg^{-1} or μg kg^{-1} in <63 μm fraction) for the period 1980–94 (metals), 1986–94 (PCBs) and 1988–94 (PAHs). Also shown is the annual average particulate load for the compounds (right vertical axis; t yr^{-1} or kg yr^{-1}).

constant of 0.1 yr^{-1} (half-life seven years). This decay rate may be considered as an upper limit, since reported half-lives for PCB degradation in sediments are usually in the order of 10 years or more (e.g. Brown *et al.* 1987, Brown and Wagner 1990, Beurskens *et al.* 1993). The sensitivity to the uncertainty in the particulate loads

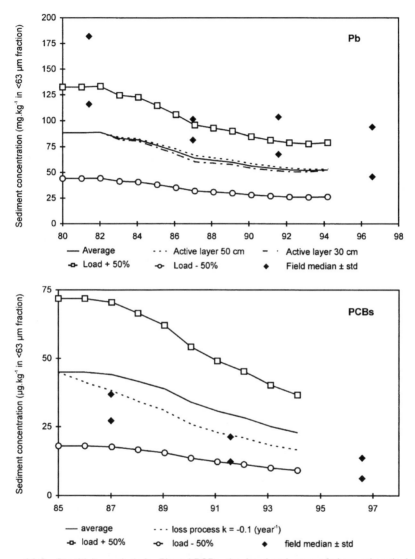

Figure 14.4 Sensitivity analysis for Pb and PCBs: simulated and range of observed particulate concentrations (mg kg^{-1} or μg kg^{-1} in <63 μm fraction) in the Dutch coastal zone for the period 1980–94 (Pb) and 1986–94 (PCBs).

was investigated by applying the earlier discussed ranges of the particulate loads of ±50% for metals and ±60% for organic compounds on Pb and PCBs respectively. The simulation results of the sensitivity analysis are compared with the range of the observed concentrations.

The simulations for Pb shows that the model is not very sensitive to increase (+10 cm) or decrease (−10 cm) of the mixing depth of 40 cm (Figure 14.4). The uncertainty of 50% in the particulate annual average load of Pb to the study area causes a much wider range in the simulated concentrations, which (partially) overlaps the range of the measured concentrations. The introduction of a loss factor of 10% per year causes a distinct reduction of the predicted PCBs concentration (Figure 14.4). However, this variation is small compared to the variation due to the range in the annual average load ($\pm 60\%$), which causes a simulation range that overlaps the field measurements.

The results for Pb and PCBs indicate that the particulate loads are the most important source of uncertainty in the model. Given this uncertainty, the model is also able to reproduce the range of the field observations for PCBs and Pb. Owing to a lack of data, this has not yet been tested for PAHs, but the results suggest that the overprediction of the sediment concentrations may be caused by an overestimation of the annual average particulate load. Hence, it may be concluded that the model can be used to predict future particulate concentrations for Pb, Cd, Cu, Zn, PCBs (and maybe PAHs) in the surface sediments of the Dutch coastal zone.

To illustrate the application of the model for the evaluation of policy measures, two simulations were made to evaluate the effect of dumping of dredged materials in the Dutch coastal zone on the Zn concentration for the period 1980–2005. The simulation of the period 1980–94 is equal to the simulation which has been discussed above. In the reference scenario the loads are kept constant at the average value of the calculated loads between 1990 and 1994. This resulted in a stabilization of the Zn concentration in the coastal sediment at the current level (Figure 14.5). In the alternative scenario the input by the dumping of dredged materials is gradually reduced by 50% between 1995 and 2000. This would yield a reduction of approximately 15% of the Zn concentrations in the surface sediments between 1995 and 2005 (Figure 14.5).

14.4 CONCLUSIONS

At first sight, the model fit for Cd, Cu and Zn is good, Pb is slightly underestimated, while the concentrations in the surface sediments for PCBs and, more pronounced, PAHs are overestimated. However, when the uncertainty in the calculated annual average particulate loads is taken into account, the model also reproduces the observations for Pb and PCBs. The delay between the annual average load reduction and the simulated concentration is about two to three years.

The results show that physical exchange of mud particles between the water column and the active bottom layer may be a dominant fate process for compounds in the non-net sedimenting, dynamic sandy sediments of the Dutch coastal zone.

The model has been used to predict changes in the concentrations of metals and organic compounds in the surface sediments of the Dutch coastal zone, depending on the changing loads from different sources, such as the dumping of dredged

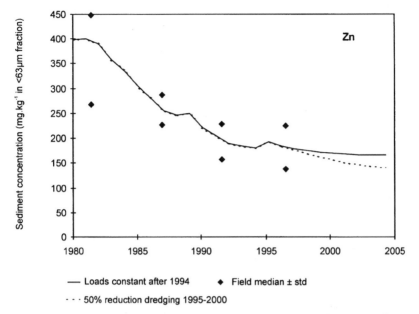

Figure 14.5 Evaluation of effect of dumping of dredged materials in the Dutch coastal zone on particulate concentrations of Zn (mg kg^{-1} in <63 μm fraction) in the surface sediments of the Dutch coastal zone during the period 1980–2005. Reference scenario: constant particulate load after 1994 at the level of the average load of 1990–94; load reduction scenario: 50% reduction of Zn load from dumping between 1995 and 2005.

materials. A reduction of the Zn load through the dumping of dredged materials by 50% between 1995 and 2000 would yield a reduction of approximately 15% of the Zn concentrations in the surface sediments between 1995 and 2005.

Sensitivity analysis has shown that incorporation of a loss process, and an increase or decrease of the mixing depth have less effect on the simulations than the uncertainty in the annual average particulate loads. This suggests that, at the present state of knowledge, the predictive value of the model would benefit more from more accurate and precise determination of the particulate loads than from more detailed model formulations (parameters and processes).

The model has great potential in meeting the declared aim of predicting changes in the concentrations of metals and organic compounds in the surface sediments in terms of changing loads from different sources. It, therefore, could be an important tool in forecasting the ecological effects of changes in these input loads into coastal waters, whether increases or decreases.

REFERENCES

Achman DR, Brownawell BJ and Zhang L (1996) Exchange of polychlorinated biphenyl's between sediment and water in the Hudson river estuary. *Estuaries*, **19**, 950–965.

Alphen JSLJ van (1990) A mud balance for Belgian–Dutch coastal waters between 1969 and 1986. *Netherlands Journal of Sea Research*, **25**, 19–30.

Anonymous (1997) *Yearbook Monitoring Data 1995: Key Figures* (in Dutch). Ministry of Transportation, Public Works and Water Management, National Institute for Coastal and Marine Management/RIKZ, The Hague, the Netherlands.

Baart AC, Berdowski JJM and van Jaarsveld JA (1995) *Calculation of Atmospheric Deposition of Contaminants on the North Sea*. Report TNO-MEP-R95/138. TNO Institute of Environmental Sciences, Energy Research and Process Innovation, Delft, the Netherlands.

Beurskens JEM, Mol GAJ, Barreveld HL, van Munster B and Winkels HJ (1993) Geochronology of priority pollutants in a sedimentation area of the Rhine river. *Environmental Toxicology and Chemistry*, **12**, 1549–1566.

Brown Jr JF, Wagner RE, Feng H, Bedard DL, Brennan MJ, Carnahan JC and May RJ (1987) Environmental dechlorination of PCBs. *Environmental Toxicology and Chemistry*, **6**, 579–593.

Brown Jr JF and Wagner RE (1990) PCB movement, dechlorination and detoxification in the Acushnet estuary. *Environmental Toxicology and Chemistry*, **9**, 1215–1233.

Eisma D (1988) An introduction to the geology of continental shelves. In *Ecosystems of the world: Continental shelves*, Postma H and Zijlstyra JJ (eds), Elsevier, Amsterdam, the Netherlands, pp. 39–92.

Gevao B, Hamilton-Taylor J, Murdoch C, Jones KC, Kelly M and Tabner BJ (1997) Depositional time trends and remobilisation of PCBs in lake sediments. *Environmental, Science and Technology*, **31**, 3274–3280.

Hegeman WJM and van der Weijden CH (1990) *Sorption Behaviour of Heavy Metals onto Rhine River Suspended Matter and Dredged Materials from Rotterdam Harbour* (in Dutch). Report, contractnumber DGW-996. University of Utrecht, the Netherlands.

Hillery BR, Simcik MF, Basu I, Hoff RM, Strachan WMJ, Burniston D, Chan CH, Brice KA, Sweet CW and Hites RA (1998) Atmospheric deposition of toxic pollutants to the Great Lakes as measured by the integrated atmospheric deposition network. *Environmental, Science and Technology*, **32**, 2216–2221.

Hoornaert S, Treiger B, van Grieken R and Laane R (1996) *Trend Analysis of the Published Concentration of Heavy Metals in Aerosols above the North Sea and the Channel for the period 1971–1994*. Report Impulse Programme in Marine Science. Belgian State Science Policy Office, contract MS/06/050, University of Antwerp, Antwerp, Belgium.

James RH, Statham PJ, Morley NH and Burton JD (1993) Aspects of the geochemistry of dissolved and particulate Cd, Cu, Ni, Co and Pb in the Dover Strait. *Oceanologica Acta*, **16**, 553–564.

Laane RWPM, E. Svendsen E, G. Radach G, G. Groeneveld G, P. Damm P, J. Pätsch J, D.S. Daniellsen DS, Føyn L, Skogen M, Ostrowski M and Kramer KJM (1996) Variability in fluxes of nutrients (N, P, Si) into the North Sea from the Atlantic Ocean and Skagerrak caused by variability in water flow. *German Journal of Hydrography*, **48**, 401–419.

Laane RWPM, Sonneveldt HLA, van der Weyden AJ, Loch JPG and Groeneveld G (1999) Trends in the spatial and temporal distribution of metals (Cd, Cu, Zn and Pb) and organic compounds (PCBs and PAHs) in Dutch coastal zone sediments from 1981–1996; possible sources and causes for Cd and PCBs. *Journal of Sea Research*, **41**, 1–17.

Lourens JM (1996) *Management of Dredged Materials, Effect of Regulation of Dumping in the Sea* (in Dutch). Report RIKZ-96.017. National Institute for Coastal and Marine Management, The Hague, the Netherlands.

Nelson ED, McConnell LL and Baker JE (1998) Diffusive exchange of gaseous polycyclic aromatic hydrocarbons and polychlorinated biphenyl's across the air–water interface of the Chesapeake Bay. *Environmental, Science and Technology*, **32**, 912–919.

Otten P (1991) Transformation, concentrations and deposition of North Sea aerosols. PhD thesis, University of Antwerp (UIA), Antwerp, Belgium.

Ras J, Klamer HJC and Laane RWPM (1996) The distribution of polychlorinated biphenyl's, polycyclic aromatic hydrocarbons and pesticides in the Straits of Dover. In *Hydrodynamics and Biochemical Processes and Fluxes in the Channel (Fluxmanche II)*, FM II Final report, MAST II, MAS2CT940089. pp. 184–194.

Reijnders PJH (1996) Developments of grey and harbour seal population in the international Wadden Sea: reorientation on management and related research. *Wadden Sea News Letters*, **1996**, 12–16.

Ruijter WPM de, van der Giessen A and Groenendijk FC (1992) Current and density structure in the Netherlands coastal zone. In *Dynamics and exchanges in estuaries and the coastal zone*, Prandle, D (ed.), Americal Geophysical Union, Coastal and Estuarine Sciences, USA.

Salomons W, Bayne BL, Duursma EK and Förstner U (eds) (1988) *Pollution of the North Sea: An Assessment*. Springer-Verlag, Berlin-Heidelberg, Germany.

Spanhof R, van Heuvel Tj and De Kok JM (1990) Fate of dredged material dumped off the Dutch shore. *Proceedings, Twenty-Second Coastal Engineering Conference, Coastal Eng. Res. Council/ASCE*. Delft, the Netherlands, July 2–6 1990, pp. 2824–2837.

Tijink J, Lourens J and Snijders G (1993) *Mass Balance for Cd in Rotterdam Harbour* (in Dutch). Report RIKZ-93.058. National Institute for Coastal and Marine Management/RIKZ, The Hague, the Netherlands.

Tijink J (1995) *Mass Balance for Heavy Metals and PAHs in Rotterdam Harbour* (in Dutch). Report RIKZ-95.023. National Institute for Coastal and Marine Management/RIKZ, The Hague, the Netherlands.

Tijink J (1998) *Micropollutants in Marine Waters Until 1996; Concentrations and Loads* (in Dutch). Report RIKZ/AB-98.119x. National Institute for Coastal and Marine Management/RIKZ, The Hague, the Netherlands.

Vethaak AD (1993) Fish disease and marine pollution. PhD thesis, University of Amsterdam, Amsterdam, the Netherlands.

Vries A. de and Klavers H (1994) Riverine fluxes of pollutants: monitoring strategy first, calculation methods second. *European Water Pollution Control*, **4**, 12–17.

Wulffraat KJ, Smit Th, Groskamp H and de Vries A (1993) *Sources of Pollutants to the North Sea* (in Dutch, with summary in English). Report DGW-93.037. National Institute for Coastal and Marine Management/RIKZ, The Hague, the Netherlands.

Zwolsman JJG (1994) *North Sea Estuaries as Filters for Contaminants*. Report T1233. Delft Hydraulics, Delft, the Netherlands.

Biomonitoring of Spatial and Temporal Patterns of Trace Metal Bioavailabilities in Tolo Harbour, Hong Kong, Using Barnacles and Mussels

GRAHAM BLACKMORE[1] AND PHILIP S. RAINBOW[2]

[1]*Department of Biology, Hong Kong University of Science and Technology,*
Clear Water Bay, Hong Kong
[2]*Department of Zoology, The Natural History Museum, London, UK*

15.1 INTRODUCTION

Evidence from biomonitors and sediments has proved useful in determining spatial patterns of trace metal pollution in Hong Kong's coastal waters. There is, however, asymmetry in both the quality and quantity of data available. Victoria Harbour (between Kowloon and Hong Kong Island) and surrounding areas (Figure 15.1) have been studied extensively and, therefore, spatial and temporal trends of metal bioavailability are well documented (Phillips and Rainbow 1988, Chan *et al.* 1990, Rainbow and Smith 1992, Blackmore 1996, Blackmore and Chan 1998). In contrast less is known about such patterns in Tolo Harbour, in particular the outer harbour. Blackmore and Chan (1998) reported a large reduction in trace metal bioavailabilities to *Tetraclita squamosa* at one Tolo Harbour site over the period 1989 to 1996. Spatial patterns are better documented with a general enrichment of metals identifiable in the inner harbour (for a review see Blackmore (1998)).

Hong Kong's entrepreneurial spirit has created a large number of light industries. It is fortunate, however, that due to a lack of developable land, heavy industry has been inhibited (Morton 1995). For this reason, all major developments in Hong Kong have been on reclaimed land and, in the past, urban and industrial centres have grown together. As new towns such as Sha Tin and Tai Po on the inner rim of Tolo Harbour developed, much of the industry previously polluting Victoria Harbour moved away from the central areas of Hong Kong into the New Territories. Prior to this development, both Sha Tin and Tai Po were small villages, and Tolo Harbour was subject to pollution from only the agricultural wastes associated with such villages (Chan *et al.* 1973, 1974). Levels of metal contamination were therefore low (Chan *et al.* 1973, 1974). Although the new developments had planned for and commissioned secondary

Forecasting the Environmental Fate and Effects of Chemicals. Edited by Philip S. Rainbow, Steve P. Hopkin and Mark Crane.

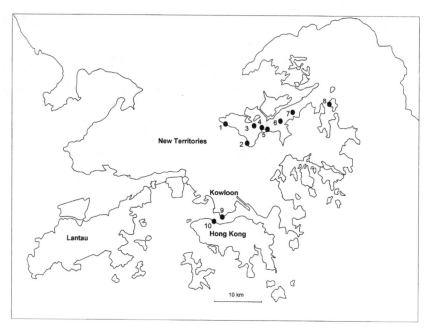

Figure 15.1 A map of eastern Hong Kong showing sampling locations. See Table 15.1 for further details.

sewage treatment plants, the long retention times in the semi-enclosed bay of the harbour have proved such measures to be inadequate. Tolo Harbour is shallow with limited water exchange (turnover 16–23 days) (Morton 1982). It has the potential, therefore, to accumulate pollutants, particularly in the inner harbour next to the new industrial towns of Sha Tin and Tai Po.

Time series of biomonitoring data provide a context for changes in metal pollution, for example in response to either increased anthropogenic input or decreases resulting from the imposition of environmental controls. Such time series, therefore, have a role to play in the future projection of the ongoing effects of changes in pollutant metal input into a habitat, and thus in forecasting. This chapter highlights Tolo Harbour, Hong Kong, as a case example to illustrate this point. The chapter concentrates on a study making use of barnacles and mussels in a programme to monitor the bioavailabilities of trace metals throughout Tolo Harbour in 1998, going on to put these data into historical context by comparison with previous but usually more limited surveys.

15.2 TOLO HARBOUR

15.2.1 STUDY AREA AND PROCEDURES USED

Details of the study locations are given in Table 15.1 and Figure 15.1. Sites 1 to 7 were in Tolo Harbour, site 8 (Tap Mun) outside the harbour, and sites 9 and 10 in

Table 15.1   Sampling sites (see Figure 15.1), details of species collected and sampling dates

Code	Site	Species collected	Date
Tolo Harbour			
1	Tai Po	*Balanus amphitrite; Perna viridis*	1/4/98
2	Ma Liu Shui	*Balanus amphitrite; Perna viridis*	10/4/98
3	Center Island	*Perna viridis*	26/4/98
4	Wu Kai Sha	*Balanus amphitrite; Perna viridis*	14/4/98
5	Starfish Bay	*Tetraclita squamosa*	14/4/98
6	Knob Reef	*Perna viridis*	26/4/98
7	Bush Reef	*Perna viridis*	26/4/98
8	Tap Mun	*Perna viridis*	26/4/98
Victoria Harbour			
9	Tsim Sha Tsui	*Perna viridis*	9/4/98
10	Queen's Pier	*Perna viridis*	9/4/98

Victoria Harbour for comparison. The barnacles sampled were *Balanus amphitrite* Darwin and *Tetraclita squamosa* Bruguière, and the mussel *Perna viridis* (L.).

Where present, barnacles and/or mussels were collected from the eulittoral zone from rocks and piers in April 1998. Samples were removed from the substratum with a stainless steel scraper, placed in individual clean polythene bags and frozen at $-20\,^{\circ}$C. At a later date, the animals were thawed, their bodies (barnacles) or soft tissues (mussels) dissected out with stainless steel instruments and rinsed quickly in distilled water before being placed in acid-washed test tubes. Ten replicate *T. squamosa* and *P. viridis* individuals were taken for analysis from each site. In the case of the smaller *B. amphitrite*, 10 bodies were pooled in each of 10 samples from each site.

Dissected bodies and soft tissues were dried to constant weight at $60\,^{\circ}$C and digested at $100\,^{\circ}$C in concentrated nitric acid (HNO_3, Aristar grade, BDH Ltd, Poole, UK). These digests were made up typically to 5 ml with deionized distilled water for subsequent analysis. Digests were analysed for arsenic, silver, cadmium, cobalt, chromium, copper, iron, manganese, nickel, lead and zinc using inductively coupled plasma optical emission spectrophotometry (ICP-OES) (Perkin Elmer Optima 3100RV). Throughout the analyses, checks were made using one aliquot of a certified reference material per 10 samples (Standard Reference Material 1566a Oyster tissue, US Department of Commerce, Technology Administration. National Institute of Standards and Technology, Gaithersburg, MD) (Table 15.2). Agreement is considered good (Table 15.2). All metal concentrations are expressed as $\mu g\,g^{-1}$ dry weight.

Metal concentrations in marine organisms are dependent on the rates of accumulation, excretion and the diluting effect of body growth. There is, therefore, the potential for an effect of size on the body metal concentration that has to be allowed for in meaningful analyses of data. Accumulated concentrations were, therefore, modelled by the power function:

$$y = ax^b$$

Table 15.2 Metal concentrations (μg g^{-1}) with 95% confidence intervals measured in samples of oyster tissue standard reference material compared against certified concentrations

	Ag	As	Cd	Co	Cr	Cu	Fe	Mn	Ni	Pb	Zn
Measured	1.44	17	4.42	0.41	1.31	62.8	467	10.7	1.87	0.315	835
CI	0.08	0.8	0.10	0.02	0.17	2.1	19	0.3	0.23	0.061	23
Certified	1.68	14	4.15	0.57	1.43	66.3	539	12.3	2.25	0.371	830
CI	0.15	1.2	0.38	0.11	0.46	4.3	15	1.5	0.44	0.014	57

where y is the metal concentration (μg g^{-1}), x the body dry weight (g), and a and b are constants. The data are, therefore, by definition, multiplicative. Data were thus transformed logarithmically to create an additive data set, essential for analysis of variance. Transformation also removes positive skews from the distribution, which are often seen in such data sets (Depledge and Bjerregaard 1990).

Data were first tested for significant slopes in the data set for each metal in each species at each site, and in the whole data set for each metal in each species (Null H_{01} is that all slope parameters equal zero). If this is not rejected (i.e. there is no size effect), a one-way analysis of variance (ANOVA) could be used to locate any differences among site means. If there is evidence to reject the null hypothesis (i.e. there is a size effect), then analysis of covariance (ANCOVA) was carried out. The data were then tested to see if the slopes for each site were equal (H_{02}). If H_{02} is rejected (i.e. not all slopes are parallel), an adjusted value of the metal concentration (for barnacles of an overall mean body weight) was used to locate differences between sites. If there was no evidence to reject H_{02} (i.e. slopes for all sites are parallel), then ANCOVA was continued to identify any differences in elevation of metal concentration/body dry weight regressions (double log) between sites. In both ANCOVA and ANOVA, *a posteriori* differences in body metal concentrations were identified using multiple comparisons of site adjusted means with Bonferroni's correction. Procedure GLM SAS® (Version 6.10: SAS Institute Inc., Cary, NC) was used to perform data analysis using methods described by Milliken (1984).

15.2.2 BIOMONITORING DATA

Weight-adjusted *Balanus amphitrite*, *Tetraclita squamosa* (bodies) or *Perna viridis* (soft tissue) metal (silver, arsenic, cadmium, chromium, cobalt, copper, iron, manganese, nickel, lead and zinc) concentrations with standard errors (which are asymmetric after antilogging) and the results of multiple comparison tests are presented. Tables 15.3 and 15.4 show weight-adjusted mean metal concentrations in *B. amphitrite* and *P. viridis*, respectively, collected during April 1998, sites being ranked in decreasing order for each metal. Tables 15.5–15.7 show weight-adjusted mean metal concentrations in

Table 15.3 Weight-adjusted mean trace metal concentrations ($\pm$ SE) in *Balanus amphitrite* from three sites within Tolo Harbour, i.e. Ma Liu Shui, Tai Po and Wu Kai Sha in April 1998. Metal concentrations of barnacles from sites sharing a common letter in the post hoc column for a particular metal are not significantly different ($p > 0.05$)

Site	μg g^{-1}	SE		Post hoc
		Lower	Upper	
Silver (Ag)				
Tai Po	2.50	1.06	5.91	A
Ma Liu Shui	2.35	1.17	4.73	A
Wu Kai Sha	0.67	0.42	1.08	B
Arsenic (As)				
Wu Kai Sha	31.0	23.9	39.5	A
Ma Liu Shui	15.9	10.9	23.2	B
Tai Po	9.82	6.18	15.6	C
Cadmium (Cd)				
Ma Liu Shui	2.43	2.18	2.71	A
Wu Kai Sha	0.95	0.87	1.04	B
Tai Po	0.69	0.61	0.78	B
Cobalt (Co)				
Ma Liu Shui	4.77	2.08	11.0	A
Tai Po	3.25	1.17	9.03	A, B
Wu Kai Sha	1.22	0.70	2.12	B
Chromium (Cr)				
Tai Po	3.32	2.08	5.28	A
Ma Liu Shui	1.68	1.14	2.49	B
Wu Kai Sha	<0.50	—	—	—
Copper (Cu)				
Ma Liu Shui	494	360	679	A
Tai Po	155	105	229	B
Wu Kai Sha	52.4	42.4	64.8	C
Iron (Fe)				
Tai Po	1470	1250	1740	A
Ma Liu Shui	404	350	466	B
Wu Kai Sha	313	277	353	B
Manganese (Mn)				
Tai Po	95.4	80.4	113	A
Ma Liu Shui	79.5	68.5	92.2	A
Wu Kai Sha	30.2	26.7	34.3	B
Nickel (Ni)				
Wu Kai Sha	8.22	3.94	17.14	A
Ma Liu Shui	2.37	0.79	7.11	B
Tai Po	1.79	0.46	6.93	B
Lead (Pb)				
Tai Po	2.51	0.84	7.54	A
Ma Liu Shui	0.53	0.22	1.3	B
Wu Kai Sha	0.40	0.22	0.72	B
Zinc (Zn)				
Tai Po	23 300	16 600	32 600	A
Ma Liu Shui	20 800	15 800	27 300	A
Wu Kai Sha	7060	5890	8 480	B

Table 15.4 Weight-adjusted mean trace metal concentrations ($\pm$ SE) in *Perna viridis* soft tissues collected in April 1998 from the sites listed in Table 15.1. Metal concentrations of mussels from sites sharing a common letter in the post hoc column for a particular metal are not significantly different($p > 0.05$)

Site	μg g^{-1}	SE		Post hoc
		Upper	Lower	
Silver (Ag)				
Tsim Sha Tsui	1.13	1.40	0.91	A
Ma Liu Shui	1.05	1.28	0.85	A
Queen's Pier	0.71	0.93	0.55	A
Center Island	0.61	0.73	0.51	A
Tai Po	0.46	0.54	0.39	A
Knob Reef	0.39	0.47	0.33	A, B
Bush Reef	0.30	0.37	0.24	A, B
Wu Kai Sha	0.14	0.17	0.12	B
Tap Mun	0.012	0.016	0.010	C
Arsenic (As)				
Tap Mun	10.4	11.0	9.91	A
Tsim Sha Tsui	9.58	10.3	8.90	A
Queen's Pier	8.32	9.08	7.62	A, B
Wu Kai Sha	6.36	6.78	5.97	B, C
Tai Po	5.49	5.80	5.19	B, C
Knob Reef	5.35	5.71	5.01	B, C
Bush Reef	4.75	5.10	4.42	B, C
Center Island	4.57	4.87	4.28	C
Ma Liu Shui	4.24	4.55	3.95	C
Cadmium (Cd)				
Tap Mun	0.55	0.59	0.52	A
Knob Reef	0.26	0.28	0.23	A, B
Queen's Pier	0.21	0.30	0.14	A, B, C
Ma Liu Shui	0.18	0.21	0.15	B, C
Tsim Sha Tsui	0.14	0.16	0.13	B, C
Center Island	0.11	0.12	0.10	C
Bush Reef	0.11	0.13	0.093	C
Wu Kai Sha	0.10	0.11	0.091	C
Tai Po	0.054	0.061	0.048	C
Cobalt (Co)				
Tap Mun	0.55	0.58	0.52	A
Tsim Sha Tsui	0.34	0.37	0.31	A, B
Wu Kai Sha	0.32	0.35	0.30	B
Knob Reef	0.32	0.35	0.30	B
Center Island	0.31	0.33	0.28	B
Queen's Pier	0.30	0.33	0.27	B
Ma Liu Shui	0.30	0.32	0.28	B
Bush Reef	0.28	0.30	0.26	B
Tai Po	0.27	0.29	0.26	B
Chromium (Cr)				
Tsim Sha Tsui	1.27	1.73	0.93	A
Queen's Pier	0.61	0.90	0.41	A, B
Tap Mun	0.32	0.40	0.00	B, C
Wu Kai Sha	0.16	0.21	0.12	B, C
Center Island	0.13	0.19	0.092	B, C
Knob Reef	0.062	0.085	0.045	B, C
Tai Po	0.052	0.071	0.037	B, C
Bush Reef	0.039	0.063	0.024	C
Copper (Cu)				
Tsim Sha Tsui	23.7	26.5	21.3	A

Continued

Table 15.4 *Continued*

Site	$\mu g\,g^{-1}$	SE		Post hoc
		Upper	Lower	
Queen's Pier	21.8	30.2	15.8	A, B
Wu Kai Sha	9.90	10.6	9.25	B
Tai Po	9.41	10.4	8.48	B, C
Ma Liu Shui	8.51	9.77	7.41	B, C
Tap Mun	7.86	8.30	7.44	B, C
Center Island	7.56	8.12	7.04	B, C
Knob Reef	6.99	7.53	6.48	C
Bush Reef	6.83	7.94	5.87	C
Iron (Fe)				
Tsim Sha Tsui	662	734	597	A
Queen's Pier	457	517	404	A, B
Wu Kai Sha	420	459	384	A, B
Tap Mun	217	234	201	B
Tai Po	194	210	180	B, C
Center Island	154	169	141	B, C
Bush Reef	130	144	118	B, C
Knob Reef	127	140	116	B, C
Ma Liu Shui	80.8	89.3	73.2	C
Manganese (Mn)				
Wu Kai Sha	44.4	48.8	40.3	A
Tai Po	32.5	35.3	29.8	A, B
Tsim Sha Tsui	28.1	31.3	25.1	A, B
Ma Liu Shui	25.7	28.6	23.2	A, B
Center Island	22.8	25.1	20.7	B
Knob Reef	22.5	24.8	20.4	B
Bush Reef	21.5	23.9	19.4	B
Tap Mun	20.3	22.0	18.7	B
Queen's Pier	19.8	22.6	17.4	B
Nickel (Ni)				
Tap Mun	3.42	4.05	2.89	A
Knob Reef	1.77	2.18	1.43	A, B
Queen's Pier	1.46	1.93	1.11	A, B
Bush Reef	0.94	1.17	0.75	A, B
Tsim Sha Tsui	0.90	1.14	0.71	A, B
Center Island	0.76	0.93	0.62	B
Wu Kai Sha	0.60	0.73	0.49	B
Ma Liu Shui	0.58	0.72	0.46	B
Tai Po	0.42	0.51	0.35	B
Lead (Pb)				
Wu Kai Sha	3.65	4.18	3.19	A
Tap Mun	2.75	3.07	2.46	A
Queen's Pier	2.69	3.24	2.23	A, B
Tsim Sha Tsui	1.77	2.07	1.52	A, B
Center Island	1.00	1.15	0.87	B
Knob Reef	0.82	0.95	0.72	B
Ma Liu Shui	0.78	0.90	0.67	B
Tai Po	0.74	0.83	0.65	B
Bush Reef	0.43	0.49	0.37	B
Zinc (Zn)				
Queen's Pier	138	185	103	A
Tap Mun	130	137	124	A
Ma Liu Shui	125	141	110	A
Center Island	99.9	107	93.7	A
Wu Kai Sha	99.0	105	93.2	A
Bush Reef	87.2	99.9	76.1	A, B
Knob Reef	85.9	92.0	80.3	A, B
Tsim Sha Tsui	84.2	92.9	76.3	B
Tai Po	79.2	87.0	72.2	B

P. viridis, *B. amphitrite* and *T. squamosa*, respectively, collected during 1986, 1989 and 1998, which are ranked in chronological order. The different weight-adjusted mean metal concentrations quoted for the same sample sets in different comparisons arise because of the different weights of the bodies/soft tissues in the whole data set compared.

15.2.2.1 Spatial variation (1998)

SILVER

Both *B. amphitrite* and *P. viridis* showed a wide range of accumulated silver concentrations. Moreover, the *post hoc* tests indicated significant differences between sites. The highest barnacle body concentrations of silver were recorded from the Tai Po and Ma Liu Shui, and significantly lower concentrations were recorded from barnacles from Wu Kai Sha (Table 15.3). In mussels, accumulated body silver concentrations were greatest in individuals from Ma Liu Shui and Tsim Sha Tsui, with significantly lower body concentrations recorded from Tap Mun (Table 15.4).

ARSENIC

Both *B. amphitrite* and *P. viridis* showed a wide range of accumulated arsenic concentrations, with significant differences between sites. The highest barnacle body concentrations of arsenic were recorded from Wu Kai Sha, and significantly lower concentrations were recorded from Tai Po (Table 15.3). In mussels, accumulated body arsenic concentrations were greatest in individuals from Tap Mun and Victoria Harbour; significantly lower body concentrations were recorded from Ma Liu Shui (Table 15.4).

CADMIUM

B. amphitrite and *P. viridis* showed a wide range of cadmium concentrations, with significant differences between sites. The highest barnacle body concentrations of cadmium were recorded from Ma Liu Shui, and significantly lower concentrations were recorded from Wu Kai Sha and Tai Po (Table 15.3). Accumulated body cadmium concentrations were greatest in mussels from Tap Mun and significantly lower concentrations were recorded from Tai Po (Table 15.4).

COBALT

Accumulated body cobalt concentrations showed little inter-site variation in both *B. amphitrite* (Table 15.3) and *P. viridis* (Table 15.4).

CHROMIUM

Both *B. amphitrite* and *P. viridis* showed a range of accumulated chromium concentrations, with significant differences between sites. The highest barnacle body concentrations of chromium were recorded from Tai Po, and significantly lower concentrations were recorded from Wu Kai Sha (Table 15.3). In mussels, accumulated body chromium concentrations were greatest in individuals collected from the Victoria Harbour sites, with significantly lower concentrations recorded from Bush Reef (Table 15.4).

COPPER

Both *B. amphitrite* and *P. viridis* showed a wide range of accumulated copper concentrations, with significant differences between sites. The highest barnacle body concentrations of copper were recorded from Ma Liu Shui, and significantly lower concentrations were recorded from Wu Kai Sha (Table 15.3). Accumulated body copper concentrations were greatest in mussels collected from the Victoria Harbour sites, with significantly lower concentrations recorded from all Tolo Harbour sites, in particular from Bush Reef (Table 15.4).

IRON

Both *B. amphitrite* and *P. viridis* showed a wide range of accumulated iron concentrations, and the post hoc tests indicated significant differences between sites. The highest barnacle body concentrations of iron were recorded from Tai Po, and significantly lower concentrations were recorded from the barnacles collected from Ma Liu Shui and Wu Kai Sha (Table 15.3). In mussels, accumulated iron concentrations were greatest in individuals collected from the Victoria Harbour sites, while significantly lower body concentrations were recorded from Ma Liu Shui (Table 15.4).

MANGANESE

B. amphitrite and *P. viridis* showed a range of accumulated manganese concentrations, with significant differences between sites. The highest barnacle body concentrations of manganese were recorded from Tai Po and Ma Liu Shui; significantly lower concentrations were recorded from Wu Kai Sha (Table 15.3). Accumulated manganese concentrations were greatest in mussels collected from Wu Kai Sha, while significantly lower concentrations were recorded from Queen's Pier (Table 15.4).

NICKEL

Both *B. amphitrite* and *P. viridis* showed a range of accumulated nickel concentrations. Moreover, the post hoc tests indicated significant differences between sites. The highest barnacle body concentrations of nickel were recorded from Wu Kai Sha, and significantly lower concentrations were recorded from Ma Liu Shui and Tai Po (Table 15.3). In mussels, accumulated nickel concentrations were greatest in individuals from Tap Mun, and significantly lower concentrations were recorded from Tai Po (Table 15.4).

LEAD

Both *B. amphitrite* and *P. viridis* showed a range of accumulated lead concentrations with significant differences between sites. The highest barnacle body concentrations

Table 15.5 A summary of metal concentrations in *Perna viridis* collected in 1986 (Phillips and Rainbow 1988, Chan *et al.* 1990) and 1998 (this study) from three sites in Tolo Harbour

Site	1986		1998	
Metal	Mean	Range	Mean	Range
Ma Liu Shui				
Ag	—	<0.05–0.1	1.05	0.45–2.92
Cd	0.2	0.17–0.23	0.18	0.09–2.59
Cr	3.7	1.7–5.7	—	<0.03
Cu	12.6	9.9–15.3	8.51	4.99–13.8
Ni	2.6	1.2–4	0.58	0.19–1.05
Pb	2.3	1.7–2.9	0.78	0.49–1.22
Zn	66	38–94	125	53–134
Tai Po				
Ag	—	<0.06–0.59	0.46	0.18–0.74
Cd	0.15	0.01–0.29	0.054	0.04–0.07
Cr	3	0–7.6	0.052	0.02–0.10
Cu	11.2	3.4–19	9.41	8.56–20.3
Ni	3.5	0.1–6.9	0.42	0.13–0.46
Pb	3.7	0–10.4	0.74	0.60–1.22
Zn	61	42–80	79.2	65.2–120.2
Wu Kai Sha				
Ag	—	<0.06-0.08	0.14	0.04–0.35
Cd	0.2	0.14–0.26	0.10	0.08–0.13
Cr	3.7	1.7–5.7	0.16	0.09–0.29
Cu	15.9	7–24.8	9.90	8.58–12.2
Ni	3.2	0–6.6	0.60	0.44–0.80
Pb	5.7	2.9–8.5	3.65	2.38–5.85
Zn	65	46–85	99.0	78.9–130

of lead were recorded from Tai Po; significantly lower concentrations were recorded in barnacles from Ma Liu Shui and Wu Kai Sha (Table 15.3). Accumulated lead concentrations were greatest in mussels from Wu Kai Sha, with significantly lower concentrations recorded from Bush Reef (Table 15.4).

ZINC

B. amphitrite showed a wide range of accumulated zinc concentrations and post hoc tests indicated significant differences between sites. The highest barnacle body zinc concentrations were recorded from Tai Po and Ma Liu Shui; significantly lower concentrations were recorded from Wu Kai Sha (Table 15.3). In contrast accumulated mussel body zinc concentrations showed limited inter-site variation (79.2–138 $\mu g\,g^{-1}$) (Table 15.4).

15.2.2.2 Temporal variation
PERNA VIRIDIS
In April 1986, Chan *et al.* (1990) and Phillips and Rainbow (1988) collected *P. viridis* from three sites in Tolo Harbour, i.e. Ma Liu Shui, Tai Po and Wu Kai Sha,

Table 15.6 Weight-adjusted mean trace metal concentrations ($\pm$SE) in *Balanus amphitrite* bodies from the sites three sites in Tolo Harbour, i.e. Ma Liu Shui, Tai Po and Wu Kai Sha. Metal concentrations of barnacles from sites sharing a common letter in the post hoc column for a particular metal are not significantly different ($p < 0.05$)

Year	$\mu g\,g^{-1}$	SE Upper	SE Lower	Post hoc
Ma Liu Shui				
Silver (Ag)				
1986	1.04	1.24	0.88	B
1989	3.01	3.47	2.61	A
1998	2.11	2.64	1.69	A
Cadmium (Cd)				
1986	2.02	2.23	1.83	C
1989	19.3	20.9	17.7	A
1998	3.25	3.68	2.88	B
Cobalt (Co)				
1989	3.35	4.51	2.50	A
1998	3.96	5.04	3.11	A
Chromium (Cr)				
1986	3.44	4.06	2.92	B
1989	11.7	13.4	10.2	A
1998	0.85	1.05	0.69	C
Copper (Cu)				
1986	109	115	103	C
1989	1810	1900	1730	A
1998	416	446	388	B
Iron (Fe)				
1986	749	903	622	A
1998	321	386	266	B
Manganese (Mn)				
1986	37.0	43.1	31.8	B
1998	78.8	91.8	67.7	A
Nickel (Ni)				
1986	10.9	12.1	9.8	A
1989	12.3	13.5	11.3	A
1998	2.67	3.02	2.36	B
Lead (Pb)				
1986	1.62	1.93	1.36	B
1989	12.6	14.3	11.1	A
1998	2.22	2.64	1.87	B
Zinc (Zn)				
1986	2990	3200	2790	B
1989	15400	16300	14600	A
1998	13600	14800	12500	A
Tai Po				
Silver (Ag)				
1986	1.80	2.14	1.52	A
1989	2.37	3.56	1.57	A
1998	3.41	5.59	2.08	A
Cadmium (Cd)				
1986	4.04	4.44	3.67	B
1989	12.4	15.7	9.76	A
1998	1.16	1.55	0.87	C

Continued

Table 15.6 *Continued*

Year	$\mu g\,g^{-1}$	SE Upper	SE Lower	Post hoc
Cobalt (Co)				
1986	2.71	4.15	1.77	A
1989	3.32	6.33	1.74	A
1998	4.17	7.21	2.41	A
Chromium (Cr)				
1986	1.18	1.69	0.83	B
1989	8.19	16.4	4.08	A
1998	2.98	5.68	1.56	A
Copper (Cu)				
1986	136	147	126	B
1989	375	457	308	A
1998	160	207	123	A, B
Iron (Fe)				
1986	710	1040	485	A
1998	1970	2890	1350	A
Manganese (Mn)				
1986	47.3	62.8	35.7	A
1998	95.2	126	71.7	A
Nickel (Ni)				
1986	10.2	13.0	8.03	A
1998	2.50	3.18	1.97	B
Lead (Pb)				
1986	4.52	5.77	3.55	B
1989	17.9	31.0	10.4	A
1998	5.31	9.70	2.91	A, B
Zinc (Zn)				
1986	4 390	4 760	4 050	C
1989	9 910	12 200	8 060	B
1998	28 200	36 900	21 500	A
Wu Kai Sha				
Silver (Ag)				
1986	1.87	2.08	1.68	A
1989	2.23	2.50	2.00	A
1998	0.58	0.64	0.52	B
Cadmium (Cd)				
1986	4.91	5.34	4.51	B
1989	16.8	18.3	15.4	A
1998	1.05	1.14	0.96	C
Cobalt (Co)				
1986	3.12	3.40	2.86	A
1989	3.86	4.19	3.55	A
1998	0.99	1.08	0.91	B
Chromium (Cr)				
1986	2.05	2.26	1.86	B
1989	7.78	8.47	7.15	A
Copper (Cu)				
1986	5225	236	214	B
1989	675	711	641	A
1998	55	58	52	C

Continued

Table 15.6 *Continued*

Year	$\mu g\,g^{-1}$	SE Upper	SE Lower	Post hoc
Iron (Fe)				
1986	1760	1990	1560	A
1998	400	445	360	B
Manganese (Mn)				
1986	78.9	90.1	69.1	A
1998	36.5	41.0	32.4	B
Nickel (Ni)				
1986	8.98	11.5	7.04	A
1989	8.57	11.3	6.53	A
1998	0.44	0.55	0.34	B
Lead (Pb)				
1986	8.23	10.3	6.57	A
1989	9.71	12.5	7.53	A
1998	1.40	1.76	1.12	B
Zinc (Zn)				
1986	4580	4800	4370	B
1989	9350	9830	8890	A
1998	8770	9190	8360	A

and their data are compared to mussels collected in the present study (Table 15.5). Cadmium, chromium, copper, nickel and lead soft tissue concentrations in *P. viridis* from all three sites all showed reductions from levels recorded in 1989. In contrast silver and zinc body concentrations increased over the period 1986–98 in *P. viridis* collected from all three sites.

BALANUS AMPHITRITE

Samples of *B. amphitrite* were collected in April 1986 (Phillips and Rainbow 1988, Chan *et al.* 1990), April 1989 (Rainbow and Smith 1992) and in the present study. These data reveal significant changes in body metal concentrations at each of the three sites, i.e. Ma Liu Shui, Tai Po and Wu Kai Sha, over the 12-year period (Table 15.6 and Figures 15.2–15.4). In general body metal concentrations were low in 1986 and were significantly greater in 1989. In 1998 metal levels had either fallen or stabilized at 1989 levels. Importantly, body concentrations of metals did not show any significant rise during the period 1989–1998.

TETRACLITA SQUAMOSA

In April 1989 Rainbow and Smith (1992) collected *T. squamosa* from one site in Tolo Harbour, Wu Kai Sha (as Starfish Bay), and the concentrations measured can be compared to those of the present study (Table 15.7 and Figure 15.5). Cadmium, copper, nickel, lead and zinc body concentrations all showed significant reductions during the period 1989–98. The largest reduction was in copper body concentrations (35.5 to 1.96 $\mu g\,g^{-1}$) and the smallest in cadmium (6.74 to 2.73 $\mu g\,g^{-1}$).

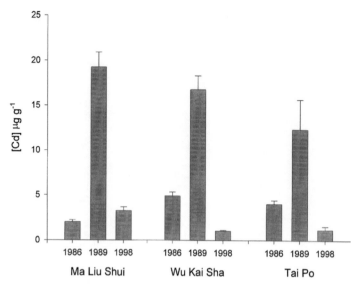

Figure 15.2   A summary of weight-adjusted mean (± 1 SE) cadmium concentrations in *Balanus amphitrite* collected from three sites within Tolo Harbour, i.e. Tai Po, Ma Liu Shui and Wu Kai Sha, during 1986, 1989 and 1998.

15.3   BIOMONITORING IN SPACE AND TIME

Biomonitors provide integrated measures of the supply of trace metals available to them, accumulating the metal taken up from all sources, such as from water and from food (Phillips and Rainbow 1993, Rainbow 1995). It follows, therefore, that different biomonitoring species in the same habitat may show slightly different patterns of metal contamination according to the different routes of uptake available to them or to any differences in the physiological handling of metals taken up by different routes (Phillips and Rainbow 1993, Rainbow 1993, 1995). Thus the use of a suite of biomonitors is to be recommended in any investigation of the metal contamination of an aquatic habitat (Phillips and Rainbow 1993, Rainbow 1995). Even closely related species may be feeding on subtly different food sources with consequently different inputs of metals for accumulation (Rainbow 1995).

   Thus it is not surprising that the mussel *Perna viridis* and the two barnacle species *Balanus amphitrite* and *Tetraclita squamosa* will not always present exactly the same picture when together at a site. Sources of trace metals to mussels and barnacles include dissolved metals and metals adsorbed on to or incorporated into suspended food particles. As a lamellibranch bivalve, *P. viridis* filters small suspended particles on its gills. Both *B. amphitrite* and *T. squamosa* feed by captorial feeding on larger particles using the three posterior pairs of thoracic limbs (cirri), and by the filtration of microscopic suspended particles (microfeeding) using the setae of the anterior cirri (Crisp and Southward 1961,

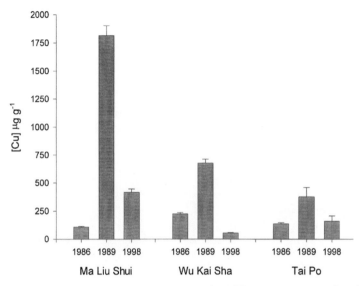

Figure 15.3   A summary of weight-adjusted mean (± 1 SE) copper concentrations in *Balanus amphitrite* collected from three sites within Tolo Harbour, i.e. Tai Po, Ma Liu Shui and Wu Kai Sha, during 1986, 1989 and 1998.

Anderson 1980, Hunt and Alexander 1991). It is probable that the two barnacles feed on slightly different size ranges of suspended particles, or at least to different degrees on different parts of the size spectrum available, with *B. amphitrite* taking more of the smaller suspended particles (Anderson 1980). *P. viridis* is probably taking relatively small particles, more akin to those taken by *B. amphitrite*.

To be an effective biomonitor of a metal, an organism must be a net accumulator of the metal concerned. This is the situation for all trace metals in barnacles (Rainbow 1987, 1998). *Perna viridis* appears to be a net accumulator of most trace metals, with a caveat on zinc for which it is a partial regulator — that is, soft tissue zinc concentrations are maintained in a relatively narrow concentration range as external availabilities increase (Chan 1988, Phillips and Rainbow 1988, 1993, Chan *et al.* 1990).

The April 1998 results for *Balanus amphitrite* and *Perna viridis* confirm that trace metal bioavailabilities in Tolo Harbour and beyond have heterogeneous distributions (Tables 15.3 and 15.4). The barnacle results show high availabilities of silver, copper, iron, manganese and zinc in inner Tolo Harbour (Table 15.3). The mussel data support these conclusions for silver and copper, although the availabilities of these two metals to *P. viridis* are higher in Victoria Harbour, as are the availabilities of chromium, iron and zinc to the mussel (Table 15.4). Examination of the two data sets suggests that Tolo Harbour did not have high availabilities of arsenic, cadmium, cobalt, chromium or nickel in 1998.

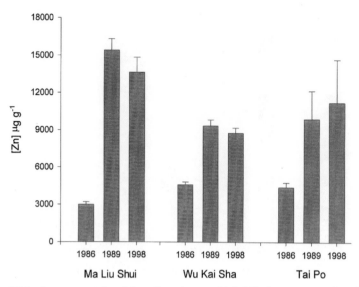

Figure 15.4 A summary of weight-adjusted mean (± 1 SE) zinc concentrations in *Balanus amphitrite* collected from three sites within Tolo Harbour, i.e. Tai Po, Ma Liu Shui and Wu Kai Sha, during 1986, 1989 and 1998.

Tables 15.5–15.7 and Figures 15.2–15.5 detail changes over time in metal availabilities in Tolo Harbour. Accumulated copper and cadmium concentrations in *Balanus amphitrite* increased from 1986 to 1989, and then decreased in the following nine years (1989–98) to approximately 1989 levels (Figures 15.2, 15.3). Accumulated zinc body concentrations in the same species of barnacle increased during the period 1986–89, but in contrast to cadmium and copper remained relatively stable until 1998 (Figure 15.4). Cadmium, copper and zinc availabilities to *Tetraclita squamosa* all fell between 1989 and 1996, with no consistent pattern between 1996 and 1998 (Table 15.7, Figure 15.5). The data for both barnacles (Tables 15.6 and 15.7) show falls in availabilities of nickel and lead from 1989 to 1998, and the *B. amphitrite* results show a similar fall in chromium availability. Manganese availability to *B. amphitrite*, on the other hand, has risen between 1986 and 1998 at Ma Liu Shui and Tai Po.

The changing patterns of metal bioavailabilities within Tolo Harbour, particularly the general decline to 1998, correlate well with the migration of local industry. As industry moved away from the central areas of Hong Kong to the New Territories (including Sha Tin and Tai Po) in the 1980s, bioavailabilities of metals increased in Tolo Harbour, reaching highs in 1989. Hong Kong industry is again moving, this time into southern China and other areas of Asia where land and labour costs are cheaper. The port of Hong Kong, however, remains efficient at transferring raw materials and trade products into and out of China. This shift is illustrated in the number of workers employed in local manufacturing industries.

Table 15.7 Weight adjusted mean trace metal concentrations ($\pm$SE) in *Tetraclita squamosa* bodies from Wu Kai Sha (Starfish Bay). Metal concentrations of barnacles from sites sharing a common letter in the post hoc column for a particular metal are not significantly different ($p < 0.05$)

Year	$\mu g \, g^{-1}$	SE		Post hoc
		Upper	Lower	
Cadmium (Cd)				
1989	6.74	5.98	7.59	A
1996	2.04	1.81	2.29	B
1998	2.73	2.4	3.05	B
Copper (Cu)				
1989	35.5	27.6	45.7	A
1996	6.96	5.97	8.11	B
1998	1.96	1.52	2.53	C
Nickel (Ni)				
1989	16.9	12.4	23.0	A
1998	1.81	1.36	2.42	B
Lead (Pb)				
1989	10.2	8.51	12.2	A
1998	3.31	2.87	3.82	B
Zinc (Zn)				
1989	10700	8830	12900	A
1996	1170	1020	1330	C
1998	5390	4570	6350	B

Such numbers have fallen from 728 000 in 1971 to 446 000 in 1994 (Morton 1996). Observed reductions in body metal concentrations of both barnacles and mussels between 1989 and 1998 indicate diminishing metal bioavailabilities as industry moves from Hong Kong.

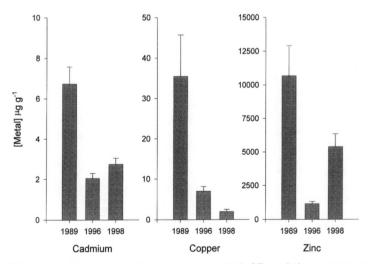

Figure 15.5 A summary of weight-adjusted mean (± 1 SE) cadmium, copper and zinc concentrations in *Tetraclita squamosa* collected during 1989, 1996 and 1998.

15.4 CONCLUSIONS

Long term biomonitoring programmes provide a cost-effective means of following changes over time in trace metal bioavailabilities, in effect of trends in the contamination of a habitat by potentially toxic agents. The recognition of any long-term trends in the availabilities of contaminants helps identify the relative contributions of different sources of contamination, and does provide forecasting potential if used with caution. As regards metal pollution in Tolo Harbour, the case study detailed in this chapter, it can be predicted from the data presented here that trace metal bioavailabilities will continue to fall and/or level out at new baselines well below the high bioavailabilities apparent in the 1980s.

ACKNOWLEDGEMENTS

Research contributing to this case study of Tolo Harbour was carried out at the 1998 International Workshop on the Marine Flora and Fauna of Hong Kong. The authors are very grateful to Professor Brian Morton of the Swire Institute for his invitation to the workshop, and for his support there.

REFERENCES

Anderson DT (1980) Cirral activity and feeding in the barnacle *Balanus perforatus* Bruguière (Balanidae), with comments on the evolution of feeding mechanisms in Thoracican Cirripedes. *Proceedings of the Royal Society London B*, **291**, 411–449.

Blackmore G (1996) Biomonitoring of heavy metal pollution in Hong Kong coastal waters, using barnacles. *Asian Marine Biology*, **13**, 1–13.

Blackmore G (1998) An overview of trace metal pollution in the coastal waters of Hong Kong. *Science of the Total Environment*, **214**, 21–48.

Blackmore G and Chan HM (1998) Heavy metal concentrations in barnacles (*Tetraclita squamosa*) in Hong Kong: a revisit. In *The Marine Biology of the South China Sea III. Proceedings of the Third International Conference on the Marine Biology of the South China Sea, Hong Kong, 1996*, Morton B (ed.), Hong Kong University Press, Hong Kong, pp. 397–410.

Chan HM (1988) Accumulation and tolerance to cadmium, copper, lead and zinc by the green mussel *Perna viridis*. *Marine Ecology Progress Series*, **48**, 295–303.

Chan HM, Rainbow PS, Phillips DJH (1990) Barnacles and mussels as monitors of trace metal bioavailability in Hong Kong waters. In *The Marine Flora and Fauna of Hong Kong and Southern China II. Proceedings of the Second International Marine Biological Workshop: The Marine Flora and Fauna of Hong Kong and Southern China, Hong Kong, 1986*, Morton B (ed.), Hong Kong University Press, Hong Kong, pp. 1239–1268.

Chan JP, Cheung MT and Li FP (1973) Marine pollution in Hong Kong. *Marine Pollution Bulletin*, **4**, 13–15.

Chan JP, Cheung MT and Li FP (1974) Trace metals in Hong Kong waters. *Marine Pollution Bulletin*, **5**, 171–174.

Crisp DJ and Southward AJ (1961) Different types of cirral activity in barnacles. *Philosophical Transactions Royal Society London B*, **243**, 271–308.

Depledge MH and Bjerregaard P (1990) Explaining variation in trace metal concentrations in selected marine invertebrates: the importance of interactions between physiological state and environmental factors. In *Phenotypic Responses and Individuality in Aquatic Ectotherms*, Aldrich JC (ed.), JAPAGA, Wicklow, Ireland, pp. 121–126.

Hunt MJ and Alexander CG (1991) Feeding mechanisms in the barnacle *Tetraclita squamosa* (Bruguière). *Journal of Experimental Marine Biology and Ecology*, **154**, 1–28.

Milliken GA (1984) SAS® Tutorial: Analysis of covariance models strategies and interpretations. *SAS Users Group International Conference Proceedings: SUGI 9*. SAS Institute Inc, Cary, NC.

Morton B (1982) An introduction to Hong Kong's marine environment with special reference to the north-eastern New Territories. In *The Marine Flora and Fauna of Hong Kong and Southern China. Proceedings of the First International Marine Biological Workshop: The Marine Flora and Fauna of Hong Kong and Southern China, Hong Kong 1980*, Morton B and Tseng CK (eds), Hong Kong University Press, Hong Kong, pp. 25–53.

Morton B (1995) Hong Kong. In *Coastal management in the Asia–Pacific region: Issues and Approaches*, Hotta K and Dutton IM (eds), Japan International Marine Science and Technology Federation, Tokyo, pp. 197–208.

Morton B (1996) Protecting Hong Kong's marine biodiversity: Present proposals, future challenges. *Environmental Conservation*, **23**, 55–65.

Phillips DJH and Rainbow PS (1988) Barnacles and mussels as biomonitors of trace elements: a comparative study. *Marine. Ecology Progress Series*, **49**, 83–93.

Phillips DJH and Rainbow PS (1993) *Biomonitoring of Aquatic Trace Contaminants*. Elsevier Applied Science, London.

Rainbow PS (1987) Heavy metals in barnacles. In *Barnacle Biology*, Southward AJ (ed.), AA Balkema, Rotterdam, pp. 405–417.

Rainbow PS (1993) Biomonitoring of marine heavy metal pollution and its application in Hong Kong waters. In *The Marine Biology of the South China Sea. Proceedings of The First International Conference on the Marine Biology of Hong Kong and the South China Sea, Hong Kong, 1990*, Morton B (ed.), Hong Kong University Press, Hong Kong, pp. 235–250.

Rainbow PS (1995) Biomonitoring of heavy metal availability in the marine environment. *Marine Pollution Bulletin*, **31**, 183–192.

Rainbow PS (1998) Phylogeny of trace metal accumulation in crustaceans. In *Metal Metabolism in Aquatic Environments*, Langston WJ and Bebianno MJ (eds), Chapman and Hall, London, pp. 285–319.

Rainbow PS and Smith BD (1992) Biomonitoring of Hong Kong coastal trace metals by barnacles, 1986-1989. In *The Marine Flora and Fauna of Hong Kong and Southern China III. Proceedings of the Fourth International Marine Biological Workshop: The Marine Flora and Fauna of Hong Kong and Southern China, Hong Kong, 1989*, Morton B (ed.), Hong Kong University Press, Hong Kong, pp. 585–597.

Author Index

Species Index

Subject Index